Home Care

A Guide to Repair and Maintenance

CMHC offers a wide range of housing-related information. For details, call 1-800-668-2642 or visit our website at www.cmhc.gc.ca

Cette publication est aussi disponible en français sous le titre : Votre maison : l'entretien et la réparation 61210

The information contained in this document represents the best available information at the time of publication. CMHC assumes no liability for any damage, injury, or expense that may be incurred or suffered as the result of the use of this publication.

National Library of Canada cataloguing in publication data

Main entry under title: Home care: a guide to repair and maintenance

Rev. ed.
Issued also in French under title: Votre maison, l'entretien et la réparation.
ISBN 978-0-660-18987-9
Cat. no. NH15-32/2003E

1. Dwellings–Maintenance and repair–Handbooks, manuals, etc.
1. Canada Mortgage and Housing Corporation.

Revised 2003
Reprinted 1985, 1988, 1990, 1992, 1994, 1995, 1997, 1998, 2001, 2003, 2004, 2005, 2006, 2007

Printed in Canada
Produced by CMHC

Contents

Keep Your Home in Good Repair

Keep Your Home in Good Repair

The home you live in represents a big investment. It's also where your family spends a lot of time, so keeping it healthy, well tended and safe is important. *Home Care* will help you keep your home in good condition and guide you through common repairs. This is not a book about renovation. It's about keeping your home in good shape, making it healthier and eliminating unsafe conditions. The best way to do this is by inspecting your home regularly and performing basic maintenance. Fixing a small problem is usually easy and prevents bigger problems that can cause serious damage and cost more money. Performing basic maintenance and repairs also provides opportunities to make your home healthier for you, your family, the community and the environment. *Home Care* tells you how to use Healthy Housing™ approaches in your home. Whether you are a renter or homeowner, *Home Care* is the book for you.

Healthy Housing™ and *Home Care*

Healthy Housing™ concept contributes to making homes that are:

- healthier for the occupants
- more energy efficient
- more resource efficient
- healthier for the global environment
- more affordable to create, operate and maintain.

Basic repairs and maintenance provide great opportunities to make better healthy housing choices. The Healthy Housing™ sections throughout *Home Care* will help you make healthier choices including:

- Occupant health—Control and repair moisture problems to avoid mold growth that can affect your health. Use materials such as water-based glues and low toxicity paints that will off-gas little or contain no harmful chemicals.
- Energy efficiency—Replace burned out light bulbs with compact fluorescent light bulbs to save on energy usage and reduce monthly electricity bills.
- Resource efficiency—Repair leaky taps promptly to conserve water.
- Environmental responsibility—Use water-based paints for any painting work. These paints are safer for disposal.
- Affordability—Maintain home heating systems to keep them operating at peak performance. This saves money as well as energy.

Home Care is Important

There are three main things to consider when caring for your home:

1. safety;
2. preventing and repairing wear and tear; and
3. maintaining indoor air quality (IAQ) and improving comfort.

Safety first

When you think about maintaining your home, safety always comes first.

- Fire safety—Prevent situations that can cause fires such as faulty heating equipment or electrical problems.
- Structural safety—Repair any problem that affects the structural safety of your home such as a damaged foundation or rotten floor.
- Indoor air quality (IAQ)—Improperly operating fuel-burning appliances can create carbon monoxide, an odorless and potentially deadly gas. This is an extremely dangerous situation and must be corrected immediately. Also, moisture problems that result in mold growth can affect the IAQ in your home and cause serious health issues.
- Occupant safety—Several elements around the home can be serious hazards. Loose handrails, damaged stairs or loose flooring are examples of details that need to be repaired to avoid dangerous tripping or falling injuries.

Fire Safety

House fires kill people

Hundreds of people die every year from house fires. Over two-thirds of all fires in Canada occur in the home. There are almost 1,000 residential and apartment fires a week, more than 50,000 a year.

What causes fires?

Carelessness and untidiness are the most frequent causes of home fires. Cigarette smoking, electrical wiring, appliances and combustible materials start fires that could have been prevented.

What can I do to keep my family safe?

- Install smoke alarms. Locate them on each level of your home (including the basement), at the top of every stairway and in hallways between the bedrooms. Test the alarm monthly by pressing the test button. If the alarm is battery-operated, change the battery on a set day, every six months. The days that the time changes between standard time and daylight saving time are easy days to remember.

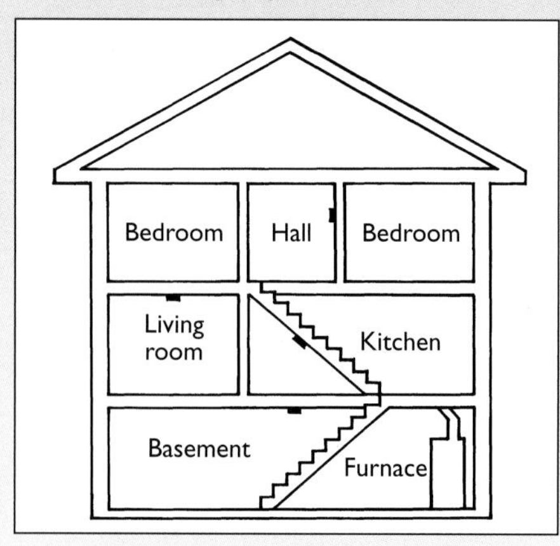

- Keep fire extinguishers handy and charged. Train everyone in your home to use a fire extinguisher properly.
- Have a family fire safety plan and practice it regularly.
- Make your home a no-smoking zone.
- If smoking is allowed in the home, always empty ashtrays in a metal ash can. Keep matches and lighters out of reach of children.
- Check that appliances are turned off before leaving your home.
- Use only appliances that are approved by the Canadian Standards Association (CSA) or Underwriters' Laboratories Canada (ULC). An approved appliance will have a label or sticker that shows it is CSA- or ULC-approved.
- Keep appliances in good working order. Replace or repair damaged plugs or cords.
- Keep stoves clean from grease that can start a serious fire.
- Keep areas around cooking and heating equipment clean and free from combustible materials. Remove curtains from windows close to the stove. Blowing curtains on an open window can easily catch on fire.
- Have heating equipment inspected and cleaned once a year.
- Clean chimneys and stove pipes at least once a year.
- Safely dispose of unneeded items or rubbish that could cause a fire.
- Keep the area under stairways clear of stored materials. A fire that starts in a stairway spreads quickly and could destroy your way out of the building.
- Store paint, gasoline and other liquids that can burn or explode in safety containers, outside and away from the house.
- Keep the grass around your home cut to prevent grass fires that can spread to your home.

Your family fire safety plan

- Sit down with your family and work out ways to get out of the house and where to meet outside.
- Review and practice the plan twice a year, for example, when the time changes between standard time and daylight saving time. These are also good times to check your smoke alarm to ensure it operates properly, and to change the battery if it has one.
- Keep all pathways and areas around outside doors clear so that you can escape quickly.
- Make sure all windows can be opened easily so they can be used to escape.
- Know whom to call if there is a fire. In many places you can call 9-1-1, but in some places you need to call the fire department. Everyone in your household should know the number to call. Post emergency numbers beside every phone in your house. If your phone has a speed dial feature, consider programming the number into it.
- Have at least one smoke alarm on each storey in your home. Most fire victims suffocate from smoke and poisonous gases. A smoke alarm could save your life once a fire has started.
- Make sure that your house number is large and easy to see from the street, especially when it is dark outside. Every second counts when help is needed. Install numbers at entrances or on both sides of a street mail box so that you can be found quickly. Reflective numbers are a good choice because they are weather resistant and can be applied on almost any surface.

How do I fight a fire in my home?

The first five minutes of a fire are critical. If a fire starts in your home, you may be able to do something when it's small and before it turns into a big fire. Each fire is different and every fire is dangerous.

Here are some helpful fire safety tips:

NEVER TAKE CHANCES. If the fire is large or too much for you to handle, get out, close the door behind you and call for help.

Clothing fires

- If clothes catch on fire, smother the fire quickly.
- Lay the victims down on the floor and roll them in a rug, coat or blanket to smother the fire while keeping their head exposed.
- Gently beat the fire out. Give the victims burn or shock first aid and get help immediately.

Cooking fires (involving fat, grease or oil)

- Turn off the stove or appliance and cover the pan or close the oven.
- Pour baking soda on the fire or use an ULC-approved, Class B fire extinguisher.
- Never use water! It will spread the flame.

Fire Safety

Electrical fires (motors, wiring and so on.)

- Unplug the appliance if possible or turn off the power.
- Use an ULC-approved, Class C fire extinguisher or pour baking soda on the fire.
- Never use water on live wiring as you may get an electric shock.

Fires in ordinary combustibles such as wood or paper

- Stay low out of heat and smoke.
- Aim the ULC-approved, Class A extinguisher at the base of the fire. For floor fires, sweep from the edges in. For wall fires, sweep from the bottom up.
- Stay outside closets and attics. Shoot the stream from the extinguisher in.

Fire Extinguishers

Fires are divided into three classes:

Class A: ordinary combustibles such as wood or paper
Class B: flammable liquids such as cooking grease or gasoline
Class C: electrical

Your fire extinguisher will have a symbol on it that shows what class of fire it can fight. Familiar yourself with the symbols and know what type of fires your extinguisher can fight. In your home, multi-purpose extinguishers that can be used for Class A, B, and C fires are usually the best types to have.

Inspect your fire extinguisher monthly. Check the gauge to ensure that the extinguisher is properly charged. Extinguishers need to be re-charged or replaced periodically. The nozzle should be clear of any obstructions and the seal should be intact. Follow maintenance instructions on the extinguisher.

Class of Fire	Extinguisher to use	Symbol
Ordinary combustibles—wood, paper, textiles, rubbish	Water, multi-purpose dry chemical or foam type Hang extinguisher on brackets supplied by the manufacturer so that the top of the extinguisher is not more than 1,500 mm (5 ft.) above the floor.	This symbol indicates that the extinguisher is for use on Class A fires. The background of the symbol will be metallic or green.
 Flammable liquids—oils grease, paints	Multi-purpose dry chemical, carbon dioxide (CO_2) or foam type Hang on brackets supplied by the manufacturer so that the top of the extinguisher is not more than 1,500 mm (5 ft) above the floor.	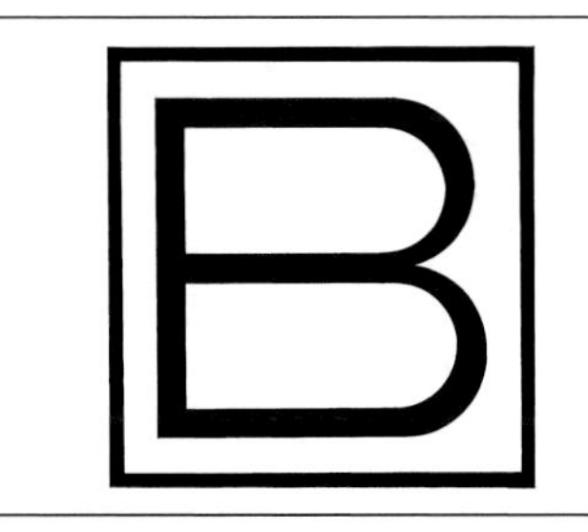 This symbol indicates the extinguisher is for use on Class B fires. The background of the symbol will be either metallic or red.
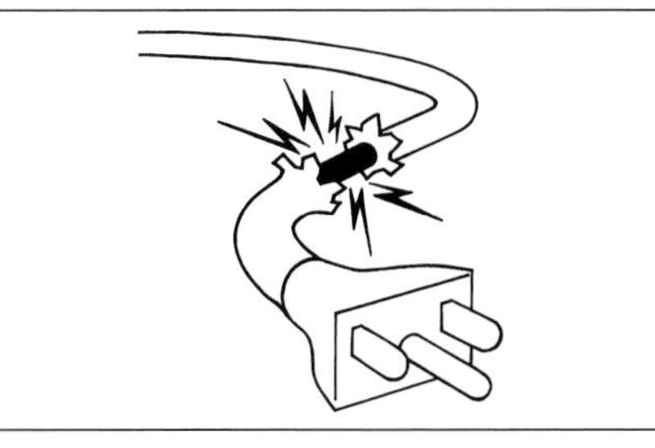 Live electric equipment, motors, wiring	Multi-purpose dry chemical or carbon dioxide (CO_2) type. Hang on brackets supplied by the manufacturer so that the top of the extinguisher is not more than 1,500 mm (5 ft) above the floor.	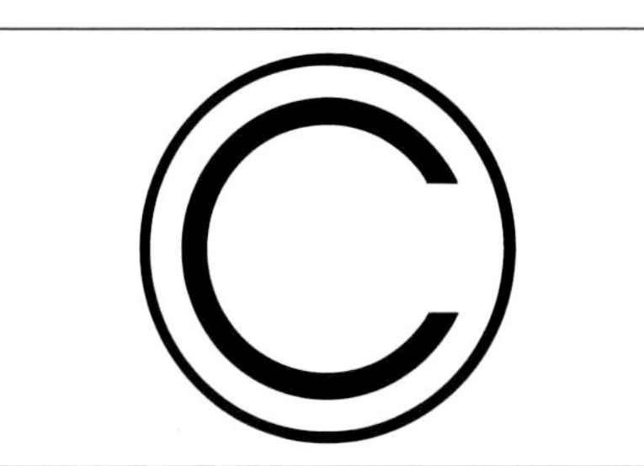 This symbol indicates the extinguisher is for use on Class C fires. The background of the symbol will be either metallic or blue.

Structural Safety

Prevent and repair wear and tear

Over time, everything wears out. A home is a busy place, inside and out, and it is exposed to a lot of wear and tear from the occupants and weather. Door hinges get loose and paint fades. The equipment inside your home and the materials used in the building also wear out or become damaged over time. Your furnace filter needs to be replaced and roof shingles may leak eventually.

Eliminate moisture problems

Moisture resulting from water leaks or excessive humidity inside your house is often the cause of serious problems. Too much humidity can cause mold growth or condensation on windows or other cold surfaces. Daily activities in your home generate a lot of moisture. Four people generate approximately nine litres (two gallons) of moisture each day from regular activities, such as bathing, cooking, drying clothes indoors, dishwashing and floor washing.

Control humidity and eliminate mold to breathe easier. Do not humidify without measuring the relative humidity level first to determine if the house is too dry. Many houses do not need added moisture. A relative humidity of 30 per cent in the winter should be sufficient to avoid breathing or mold problems. Relative humidity above 50 per cent next to cold surfaces can lead to mold growth. (Note: The relative humidity next to cold surfaces is higher than that in the middle of the room)

Maintain indoor air quality and increase comfort

We spend up to 90 per cent of our time indoors—most of it in our homes. The more you can minimize and control moisture and contaminants in your home, the more comfortable you'll be and the more likely that your home will have better indoor air quality. Controlling moisture is critical for avoiding mold growth and dust mites that affect the air quality and comfort in your home. Other things can also affect IAQ and even cause health problems. There are two categories of contaminants—biological and chemical.

Biological contaminants include molds, dust mites, pollen, animal dander and bacteria.

Chemical contaminants include cigarette smoke, combustion gases from fuel-burning appliances, and emissions from cleaning products, furnishings, building materials and hobbies.

An effective IAQ strategy includes:

1. **Eliminate.** Eliminating the cause of the problem is always the best option. For example, molds need moisture and foods to grow such as paper, cardboard or anything organic. By properly maintaining the humidity levels in your home, molds will be unable to grow. You can also ban all cigarette smoking from indoors, keep animals outside all the time, and use non-toxic and fragrance-free cleaning products.
2. **Separate.** Keeping separate what you can't eliminate is another option. For example, furnishings made with particleboard can emit gases that will affect the indoor air. Seal the particleboard with a water-based urethane. Keep cleaning and hobby products isolated in sealed plastic bins.
3. **Ventilate.** Mechanical ventilation will help to reduce moisture and dilute the levels of pollutants in your home.

Know Your Limits

Home Care covers the more routine repair and maintenance situations. Some repairs are more complex than others and may be better done by a professional. Sometimes maintenance tasks may seem straightforward, but will include safety checks that only a well-equipped professional is capable of doing. For instance, annual servicing of gas or wood heating appliances and chimney cleaning are opportunities for a professional to spot unsafe conditions. The work must be completed by a licensed or qualified person.

Go ahead with only those repairs that you feel you are capable of doing. You may live in a rental unit and not be permitted to do certain repairs. Doing repairs yourself can save you money, but trying to do a repair that is beyond your skills may end up costing you or your landlord more. Check with your landlord before making any repairs.

If in doubt about how to fix a problem—call a professional.

If in doubt about whether you are permitted to do a repair—call your landlord or the applicable utility company.

Home Care provides an estimate of the skill level required to carry out each maintenance or repair task.

Skill level rating: 1 - Simple maintenance—no previous experience or training required

Skill level rating: 2 - Handy homeowner—some experience in using tools and doing repairs required

Skill level rating: 3 - Skilled homeowner—skills and experience in using tools and doing more advanced repairs required

Skill level rating: 4 - Qualified tradesperson/contractor—training, skills and experience required

Skill level rating: 5 - Specialist/Expert—specialized training, advanced skills and experience required

About *Home Care*

- In *Getting Started*, you'll find out about assessing your home and about the basic tools you'll need.
- *Basic Home Repairs and Maintenance* explains common situations for all the major areas of your home. Each section offers tips on preventing problems and making repairs easier. There are also handy reminders about safety and pointers on making your home healthier by choosing Healthy Housing™ improvements that can improve occupant's health, energy efficiency, resource efficiency, environmental responsibility and affordability.
- There will be times when you will need a professional either to do the work or help you to figure out what needs to be done. In *Getting More Help*, you can learn about hiring the professionals you need. You'll also find a list of excellent publications and other resources to help you learn more.

Getting Started

Getting Started

Assessing Your Home

The first step in doing maintenance and repairs is to survey your home to see what needs to be done. Start outside and assess the foundation, exterior walls and roof. Look for cracks in the foundation, damaged or stained siding, cracks around doors and windows, and damaged or missing shingles. When you are finished outside, move inside. Begin in the basement and move through the entire house looking at the living areas and the attic. Look for moisture stains, bubbled paint, plumbing leaks, cracks in walls and hazardous conditions.

As you do the inspection, note all the repairs that need to be done. Consult the *Homeowner's Inspection Checklist* for more complete home inspection information.

Homeowner's Inspection Checklist (62114) will help you make sure your home is safer; more energy-efficient and more comfortable all year round, in as little as a few minutes a week. The practical guide has "how-to" tips for every room of your house on the most common problems to look for, the most effective solutions, plus the handy Healthy Housing Evaluation Tool and Basement-to-Roof Maintenance Calendar. To order, contact the nearest CMHC office or call 1-800-668-2642.

Once you've completed your inspection, set your priorities.

1. Safety first. Any problems that relate to safety need to be done immediately. Examples are a damaged chimney that could be a fire hazard, or a broken stair that could cause someone to fall.
2. Small problems that could become big problems. Repairs that will lead to a major or more costly repair should have a high priority. A leaking pipe could lead to damage to the surrounding wall and serious mold growth.
3. Maintaining and improving comfort. Any repairs that maintain or improve comfort can be beneficial. Regular cleaning or replacing of furnace filters can keep the air cleaner and will help to keep the furnace operating well.
4. Other repairs needed such as touching up paint and oiling squeaky door hinges.

There are also some items that should be checked at particular times of the year. Here is a simple, monthly maintenance checklist to help you.

January
- ✔ Check furnace and air exchanger filters. Clean or replace filters when they are dirty.
- ✔ Inspect the house for excessive moisture.

February
- ✔ Check furnace and air exchanger filters. Clean or replace filters when they are dirty.
- ✔ Inspect plumbing for drips and leaks.

March
- ✔ Check furnace and air exchanger filters. Clean or replace filters when they are dirty.
- ✔ Inspect the home for moisture damage.
- ✔ Inspect the home for interior maintenance.

April
- ✔ Check furnace and air exchanger filters. Clean or replace filters when they are dirty.
- ✔ Test the smoke detector and replace the battery.
- ✔ Check the fire extinguisher pressure gauge. Get extinguisher re-charged if needed.
- ✔ Inspect the basement for signs of water leakage.
- ✔ Check the siding and outside of your home for winter damage.
- ✔ Clean any debris from the eavestroughs and downspouts. Reattach any sections that are loose.
- ✔ Inspect the grade and landscaping for proper drainage.

May
- ✔ Inspect windows and doors for operation and screens for needed repairs.
- ✔ Inspect foundation walls for cracks and leaks.
- ✔ Check furnace and air exchanger filters. Clean or replace filters when they are dirty.
- ✔ Clean the chimney for any wood-burning appliance at the end of the heating season.

June
- ✔ Check furnace and air exchanger filters. Clean or replace filters when they are dirty.
- ✔ Have the septic tank checked and cleaned, if needed (usually every three years).
- ✔ Inspect the condition of the roof for loose or missing shingles.
- ✔ Check the yard for exterior maintenance needs such as fences and shed repairs or tree and bush trimming.

July
- ✔ Check furnace and air exchanger filters. Clean or replace filters when they are dirty.
- ✔ Check drainage ditches for debris and clean, if needed.
- ✔ Check the home for interior maintenance.

August
- ✔ Check furnace and air exchanger filters. Clean or replace filters when they are dirty.
- ✔ Check the home for exterior maintenance.

September
- ✔ Check furnace and air exchanger filters. Clean or replace filters when they are dirty.
- ✔ Clean the chimney and have the furnace serviced.
- ✔ Vacuum electric heaters to remove dust.

October
- ✔ Test the smoke detector and replace the battery.
- ✔ Check the fire extinguisher pressure gauge.
- ✔ Check furnace and air exchanger filters. Clean or replace filters when they are dirty.

November
- ✔ Inspect your home for excessive moisture.
- ✔ Check furnace and air exchanger filters. Clean or replace filters when they are dirty.
- ✔ Check the home for interior maintenance needs.

December
- ✔ Check furnace and air exchanger filters. Clean or replace filters when they are dirty.
- ✔ Check windows and doors for ice build-up.
- ✔ Check electrical cords, outlets and plugs for damage.

Professional home inspectors can also help to inspect your home or a prospective home. Houses with significant areas of mold may need to be inspected by a knowledgeable indoor air quality investigator. For suggestions on hiring a home inspector or indoor air quality investigator, refer to the *Getting More Help* section in this publication.

Your Tool Kit

Once you've taken a survey of what needs to be done and decided what to do first, you'll want to get to work.

You'll need some basic tools to do repair and maintenance jobs around the home. Choosing tools is very personal. Select tools that feel "right" in your hand. Many tools have special grips to make it easier to hold onto the handle. Tools are also available in lighter weights or special sizes that may work better for people with smaller hands. Most of the tools described in this section can be bought one-by-one, as you need them. This listing is not meant to be a complete list of all the tools you might want to own. More expensive tools or other tools you might need can be rented or borrowed.

Basic tools

- caulking gun
- chisel—25 mm (1 in.) for wood
- combination square
- drill and drill bits—cordless is handiest
- drop cloth
- dust mask
- fasteners—a basic selection of screws, nuts, and bolts
- flashlight
- hammer—454 g (16 oz.)
- ladders—step and extension
- level
- nail set
- pliers—locking, needle-nose, slip-joint
- retractable steel measuring tape—5 m (16 ft.)
- safety goggles
- sandpaper (various grits)
- saws—cross cut and hack saw
- scrapers—various widths
- screwdrivers—various sizes in square drive (Robertson), star (Phillips) and slotted type
- socket set and driver
- staple gun
- stiff brush or wire brush
- toilet plunger
- utility knife
- wrench—adjustable

Useful extras

- chalk line
- clamps
- cold chisel
- drain auger
- ear protection for using power tools
- framing square
- knee pads
- mallet
- mitre box
- multipurpose tool
- plane—block type
- saws—bow, backsaw and circular saw
- stud finder
- wrecking bar—flat
- wrench—hex, pipe

Chisels

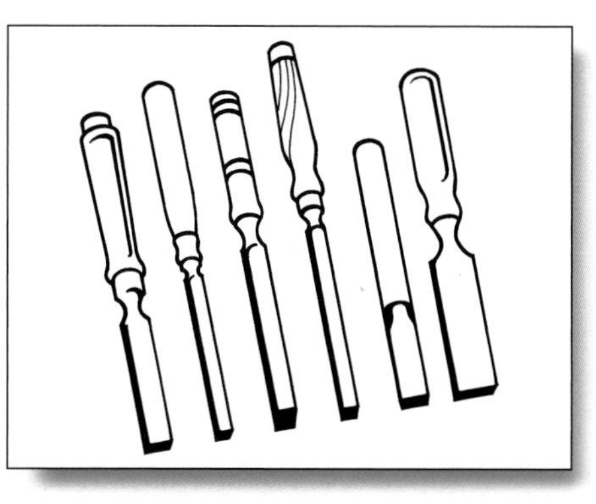

A *chisel* is a sharp, cutting and refining tool with a bevelled cutting edge. *Wood chisels* come in standard sizes and are used to chip, cut, pare and shape wood. They are designed to be palm- or mallet-driven. They have plastic or wood handles. Use a honing stone to sharpen wood chisels.

Cold chisels are usually made of solid steel and are used to chip out loose masonry or to cut sheet metal. Use a grinding wheel to sharpen a cold chisel.

Drill and drill bits

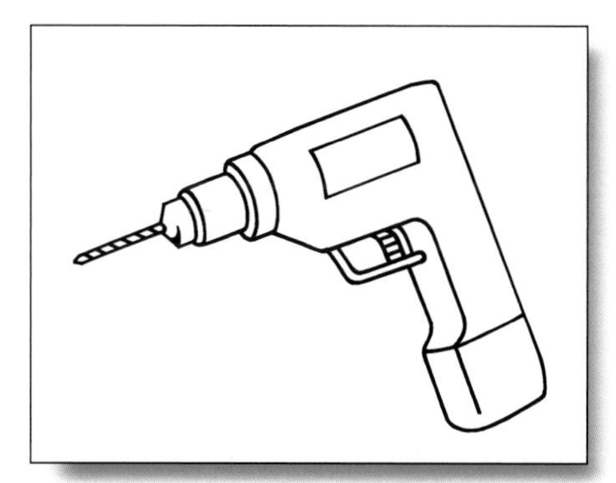

You'll need a *drill and drill bits* to make holes for bolts, screws, other fasteners and for other purposes. Cordless, battery-operated drills are the handiest types for work around the house. Heavy-duty cordless drills and electric drills are also available for jobs that require a lot of drilling. The most popular light-duty drills take bits up to 9 mm (3/8 in.) in diameter. Heavier duty drills may take bits up to 12 mm (1/2 in.) in diameter or larger. Drill bits are available in a variety of shapes, sizes and serve many purposes including standard drilling and specialty uses. When selecting the bit, choose one that is sized and designed for the type of drilling that you plan to do. Install the bit according to the drill manufacturer's directions.

Fasteners

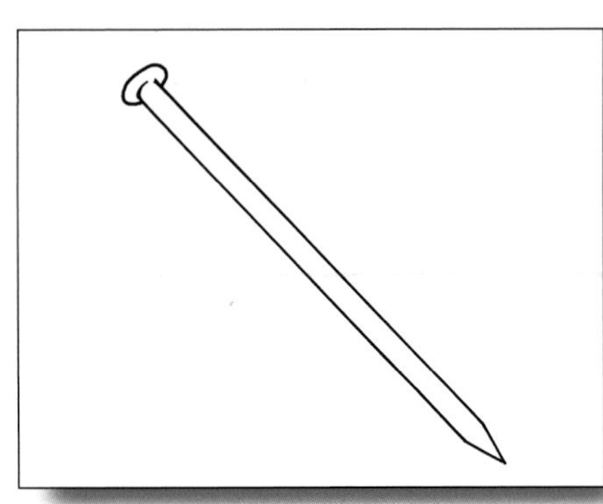

Common nails have large heads. Use them for rough work where appearance is not important.

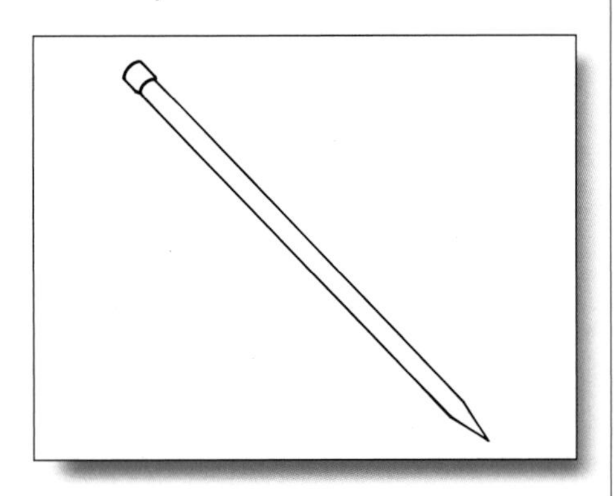

Finishing nails have small heads. You can drive them below the surface with a nail set and cover them with filler. Use them where appearance is important and you do not want the nail to be visible.

Special nails are available for a variety of uses (drywall, roofing and masonry) or added strength (spiral nails).

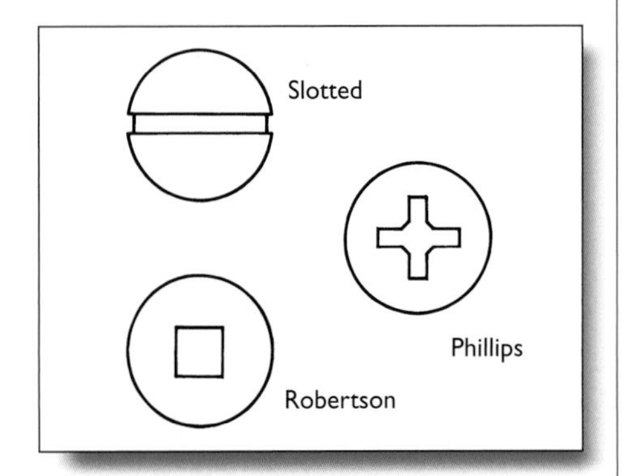

Use *screws* where holding strength is important. Screws come in various types for different jobs such as general purpose, drywall and sheet metal work. The most common drive heads for screws are the square-drive (Robertson), star (Phillips) and slotted.

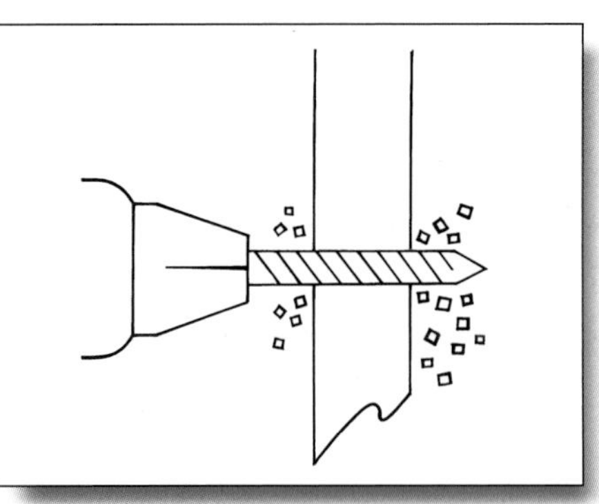

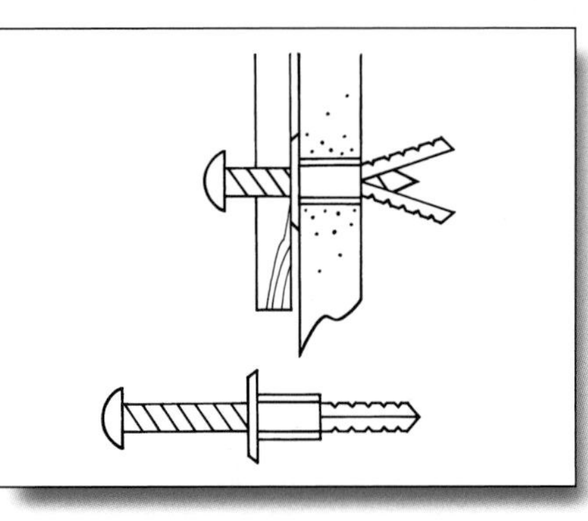

Plastic anchor screws are increasingly being used for attaching light items to drywall. First, drill a small hole in the wall and drive the casing even with the wall surface. Put the screw through the item and into the casing. Tighten the screw. Some self-tapping anchor screws can be inserted into drywall by lightly tapping and then screwing them into place.

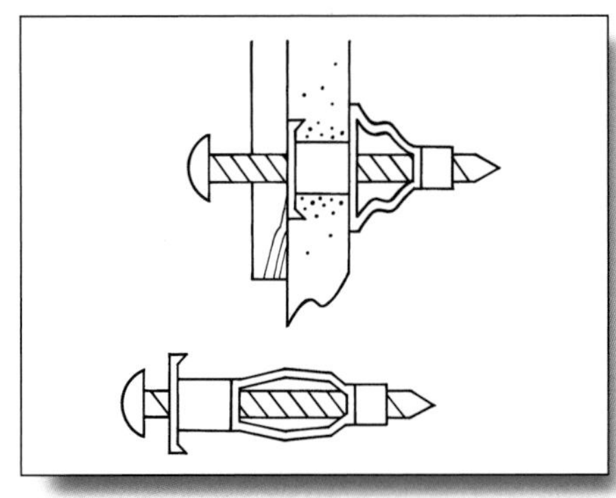

Use *molly bolts* or *toggle bolts* to attach heavy items to drywall-finished walls.

Molly bolts have two parts, a bolt and a casing. Drill a small hole in the plaster and drive the casing in even with the wall surface. Tighten the bolt to spread the casing in the back. Remove the bolt and put it through the item you are hanging, then into the casing. Tighten the bolt.

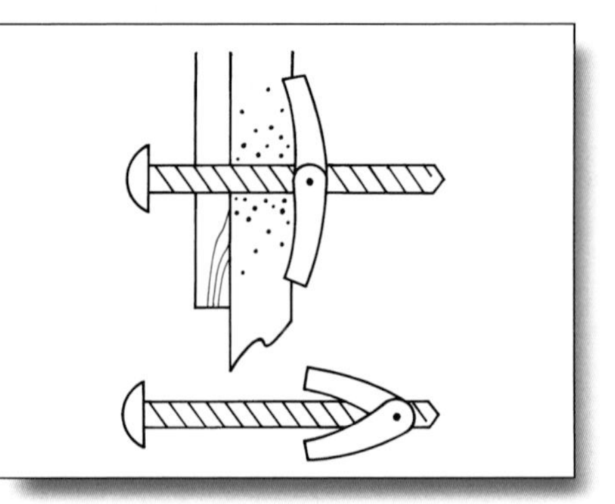

For a *toggle bolt*, drill a hole in the plaster large enough for the folded toggle to go through.

Remove the toggle. Put the bolt through the item you are hanging. Replace the toggle. Push it through the wall and tighten with a screwdriver.

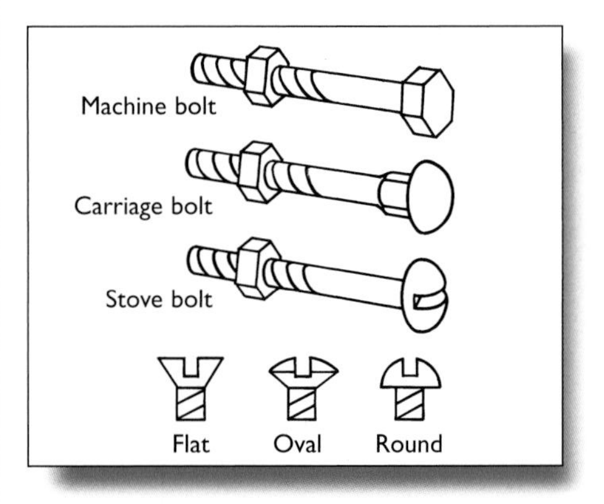

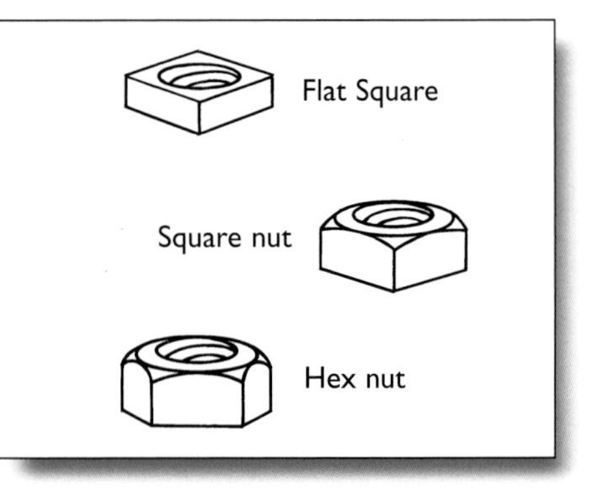

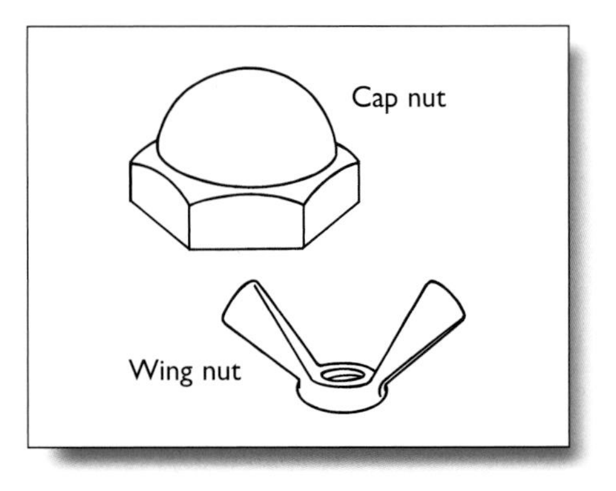

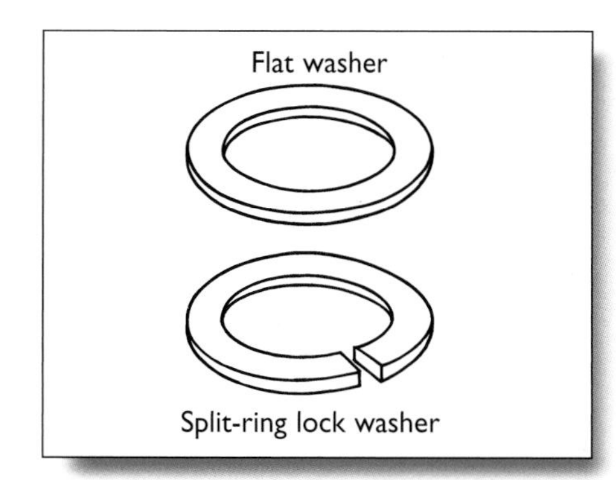

Use *nuts, bolt,* and *washers* to fasten heavy items together or when extra rigidity is required. Use one or two washers on either side of the item to prevent the nut or bolt head from slipping or digging into the item. Drill a hole for the bolt in the two items to be attached. Slip one washer over the bolt and thread the bolt through the hole. Put the second washer over the exposed threaded end of the bolt and thread on the nut. Tighten the nut with a wrench while holding the bolt head in place with a screwdriver or wrench.

Use lock washers in situations where vibration tends to loosen nuts. The lock washer always goes between the item (or flat washer) and the nut.

Hammers and mallets

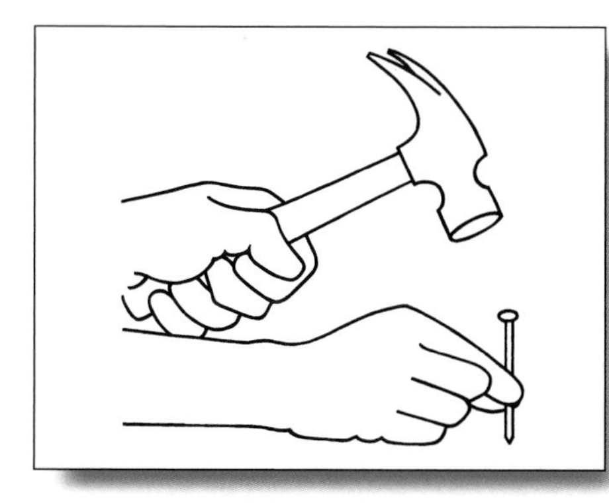

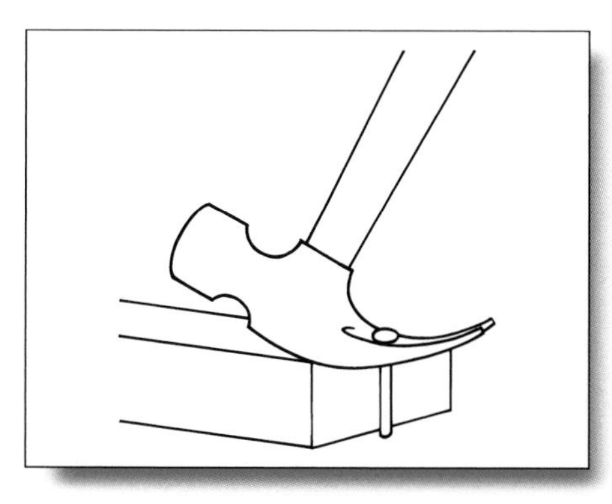

A 454-g (16 oz.) *claw hammer* is a good general purpose hammer. Hold the hammer near the end of the handle for more hitting power. To start a nail, hold the nail in place and tap it gently a few times until it is set firmly, then hit it straight in.

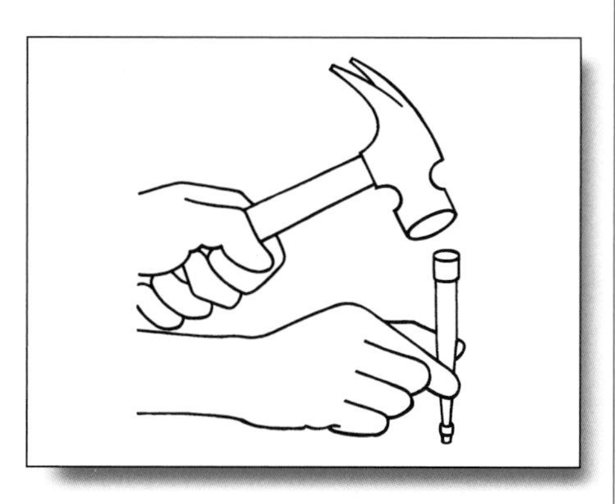

To remove a nail, use the claw end of the hammer. Place a small block of wood under the hammer head to provide extra leverage and avoid marking the surface.

Use a wood or rubber headed *mallet* to drive a wood chisel and to shape metal.

Ladders

An extension ladder and a stepladder are often needed for home repairs. Proper placement of all ladders is important for safety. Watch out for power lines when placing ladders. Contact with power lines can cause serious injury or death.

An *extension* ladder is needed to work at heights above 2.5 m (10 ft.). A *stepladder* is self-supporting and very versatile. Do not use a stepladder over 2.5 m (10 ft.) high because it will be too unstable. Choose an extension ladder instead.

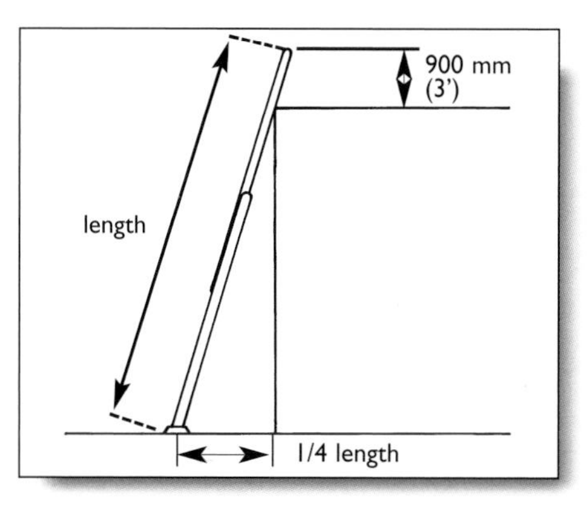

The base of the ladder should be placed so that the distance from the wall is equal to one-fourth the length of the ladder.

Here is a simple method to correctly place a ladder:

- Lean the ladder against the wall.
- Place your feet at the base of the ladder.
- Stretch your arms out, straight in front of you. If your arms comfortably reach the rungs, the ladder is at the correct angle. If your arms do not reach comfortably, adjust the angle of the ladder.

Ladder safety

- Use a ladder with a non-slip base if it is standing on smooth or sloping surfaces.
- Never let the ladder rest against windowpanes or glass doors.
- Check the treads and side rails to make sure that they are sound and tight.
- Is the ladder long enough? At least two treads, or 900 mm (about 3 ft.) should extend above where you need to climb.
- Having a helper hold the base of an extension ladder or tying the ladder off at the top are good practices.
- Some work, such as painting a high wall, may be unsafe to do from a ladder. Use scaffolding.
- Always face the ladder and use both hands when climbing up or coming down. Raise or lower your materials and tools with a **rope or sling.**
- Never lean from a ladder. If something is beyond safe and easy reach, move the ladder.
- Never leave the ladder standing except for short breaks during your work. When you finish your workday, take the ladder down and put it away.

Levels

A level helps you to make things horizontally level or vertically plumb. A *carpenter's level* is a straight bar that has several vials of fluid positioned in the bar. The vials have a bubble inside and marks showing the centre of the vial. To test whether something is level or plumb, hold the bar against the surface you're checking and see whether the bubble stops moving in between the centre marks of the vial. If it does, the surface is level or plumb.

Mitre box

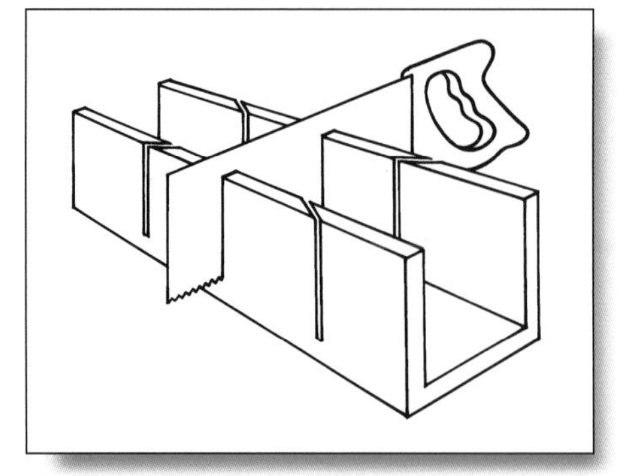

A mitre box is a simple guide to use when sawing moulding or trim at exact angles. Inexpensive mitre boxes made of wood are used with a separate backsaw.

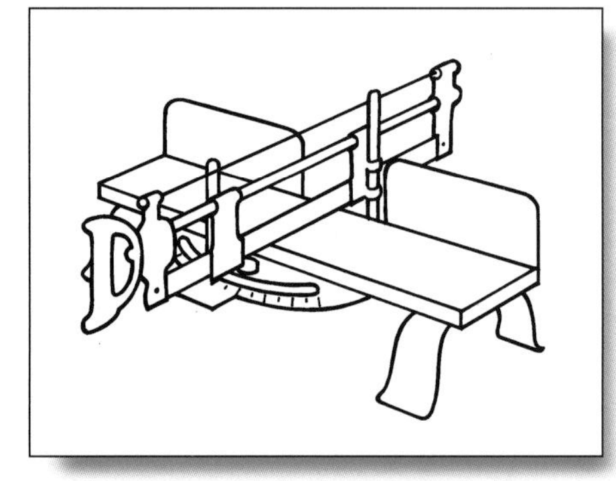

More expensive ones made of steel, come with a backsaw set in a steel box. Electric mitre saws are a popular choice at a higher price. They can often be rented. Power mitre saws are very handy since they are usually very accurate, easy to use and able to cut compound angles.

Nail sets

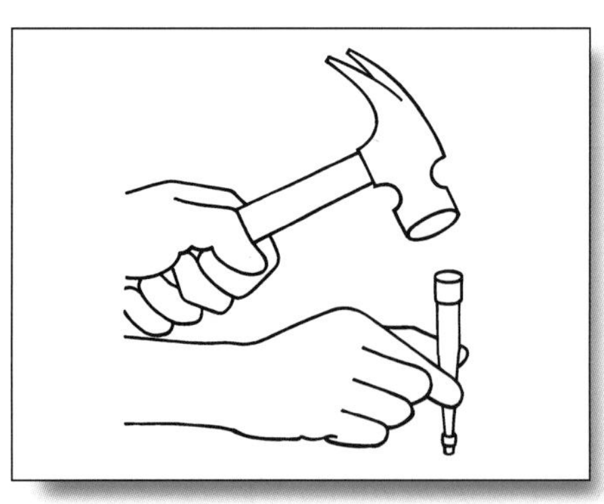

A nail set is a small metal device used to sink the heads of nails slightly below the surface of the wood without harming the finish.

Planes, rasps and files

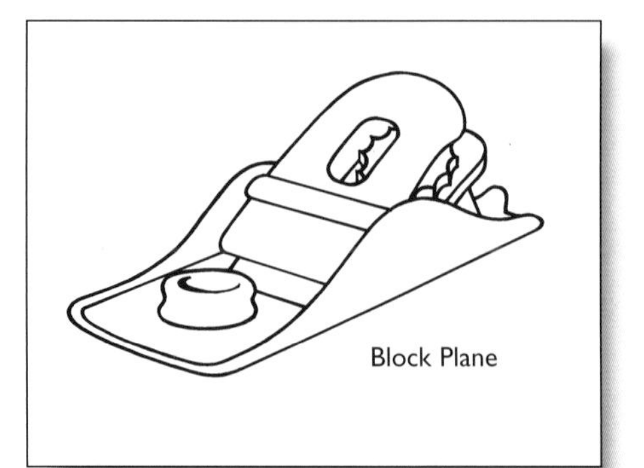

Planes, rasps and *files* are used to remove excess material and smooth surfaces. A *block plane* is handy to trim down the edge of a door that is sticking.

Pliers

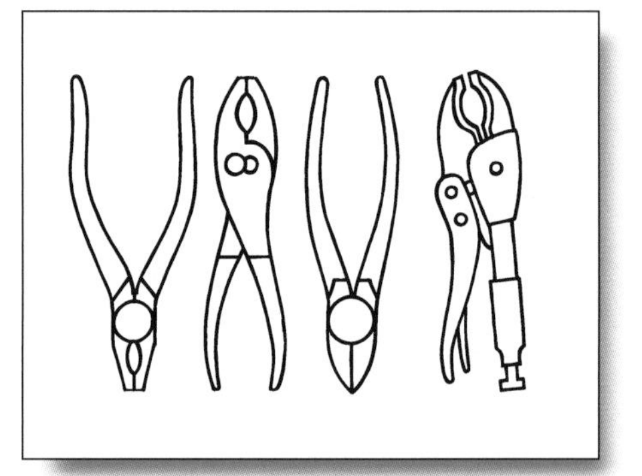

There are four basic types of pliers—gripping, adjustable, cutting and locking. For your tool kit, you'll probably want at least one of each type. *Slip-joint pliers* are adjustable and can be used for many jobs around the house. Use them to help remove nails or brads since they are small and do not have a head to grab with a claw hammer. Pull the nail out at the same angle as it was driven in. Use a small block of wood under the pliers for leverage and to protect the finish. *Groove-joint pliers* are also adjustable and can grip objects of any shape.

Long-nosed pliers are a gripping type and good for bending wire, especially for electrical work. Do not use pliers to turn nuts because they will damage them. Use a wrench instead.

Use *diagonal cutting pliers* when you need to cut wire.

Locking pliers have an adjusting knob that enables you to clamp the pliers' jaws tight. Once you're done with your task, release the lever to unlock the pliers.

Pry bars

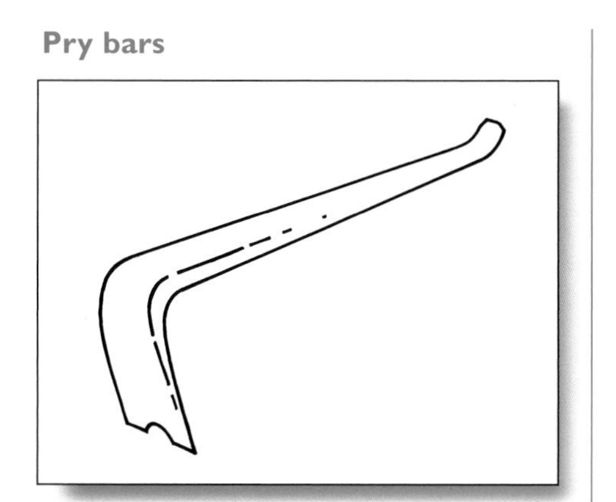

Pry bars or wrecking bars are made of steel and come in several sizes. They usually have a claw at one end for removing nails and a wedge-shaped prying edge on the other end. Use a small pry bar made of flat steel stock to remove mouldings. Use a medium-sized bar about 600 mm (2 ft.) long, made of hexagonal stock for heavier carpentry work.

Safety equipment

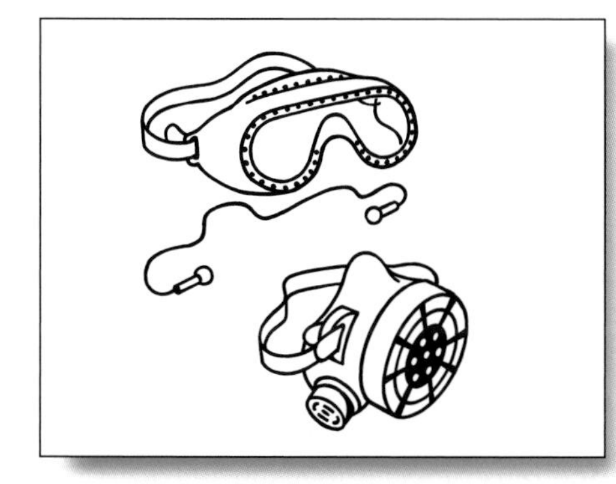

Wear a *dust mask* to protect yourself when working with drywall, wood or fibrous insulation. Refer to the ratings on the packaging and choose a mask that will protect you for the specific job you're doing.

Proper *ear protection* while using power equipment is essential. Refer to the safety ratings on the packaging and choose the ear protector that is rated for the specific job you'll be doing.

Goggles will protect your eyes from debris, drywall dust and wood chips as you work. Goggles are rated for safety and for the specific job. Choose goggles that meet your needs and provide the proper protection.

Saws

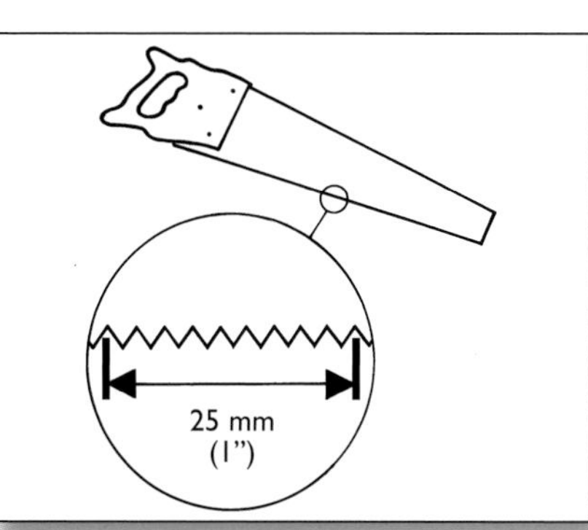

A *handsaw* or *crosscut saw* with about 10 teeth per 25 mm (1 in.) is good for most household work.

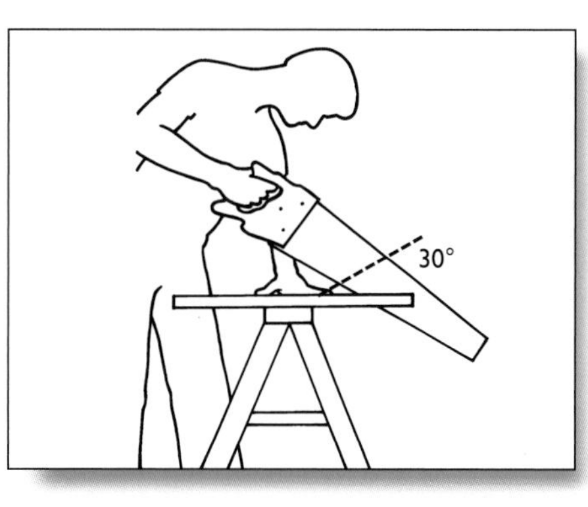

Mark your cut lines clearly. Support your work and hold it firmly near where you are cutting. Pull the saw back several times to start a groove. Saws cut on the forward stroke. Let the weight of the saw do the cutting.

A *backsaw* has fine teeth on one edge and reinforcement for rigidity along the back edge. Use a *backsaw* with a *mitre box* for precise cuts.

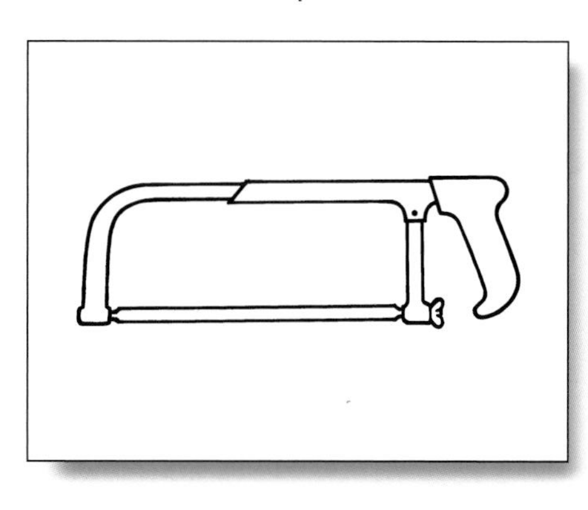

Use a *hacksaw* with a removable blade to cut metal such as bolts, nuts and pipes.

A *portable circular* saw can save you lots of muscle power and time. Use it as a crosscut saw to cut across the grain of the wood or a ripsaw to cut along the grain of the wood.

Adjust the saw blade so that the amount of blade extending below the "shoe" is one tooth deeper than the thickness of the material to be cut. As you guide the saw forward, the blade safety guard is automatically pushed back, exposing the blade for cutting

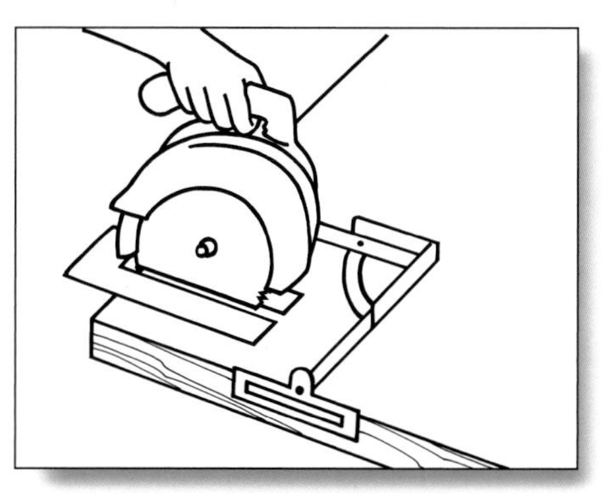

For ripping work (cutting the board along the grain), use a *ripping guide*. After adjusting the blade, set the ripping guide the same distance from the saw blade as the width of the material to be cut off. Place the guide against the edge of the piece as you cut.

Circular saw safety

- Make sure the saw has a guard that will automatically adjust while using it to keep the saw teeth from being needlessly exposed.
- Make sure that power to the saw cuts off when the trigger is released.
- Always wear goggles or a face mask to prevent injury.
- Examine the material that you are going to cut. Make sure it is free of nails or other metal before you begin.
- Ensure the material is supported properly so that it won't move or "kick" as you cut.
- Check that the area under the cut line is clear to allow the blade room without cutting into the support.
- Hold the saw firmly against the work when cutting.

- Never feed a circular saw into the material backwards. It will suddenly grab the material and shoot it toward you.
- Never overload the saw motor by pushing too hard or cutting material that is too thick for the saw.
- Always try to make a straight cut to avoid binding the saw blade. If the blade binds or sticks in the groove, back the saw out slowly and firmly in a straight line. As you continue cutting, adjust the direction of the cut so that you are cutting in a straight line.
- A sharp blade makes cutting much easier and does not overload the saw.
- Always pull the electric plug before adjusting the saw or inspecting the blade.

Scrapers

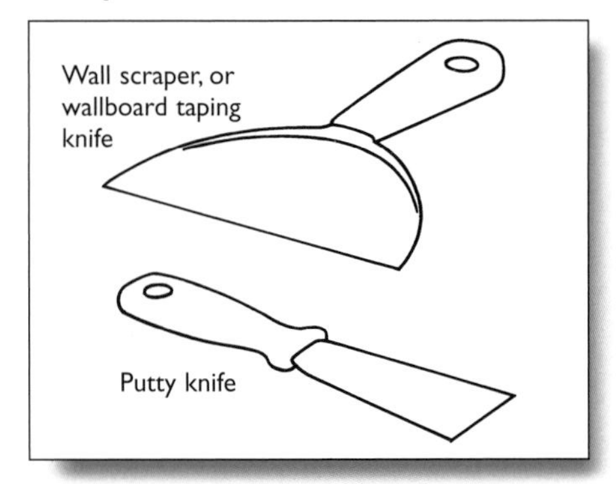

Scrapers have thin, flexible, broad blades that make them useful for removing old wallpaper, loose paint and plaster, and for filling holes and cracks with new plaster. They come in various sizes.

Screwdrivers

You need three types of screwdrivers in assorted sizes for home repairs: the *slotted*, the *star type (Phillips)* and the square drive *(Robertson)*. The blade of the screwdriver should snugly fit the opening in the screw head. When using a screwdriver, push against the head of the screw as you turn it.

It's easier to screw into wood if you make a small test hole first with a nail or drill to avoid splitting the wood. It will also be easier to screw into the wood if you rub soap or wax on the screw threads.

Squares and tape measures

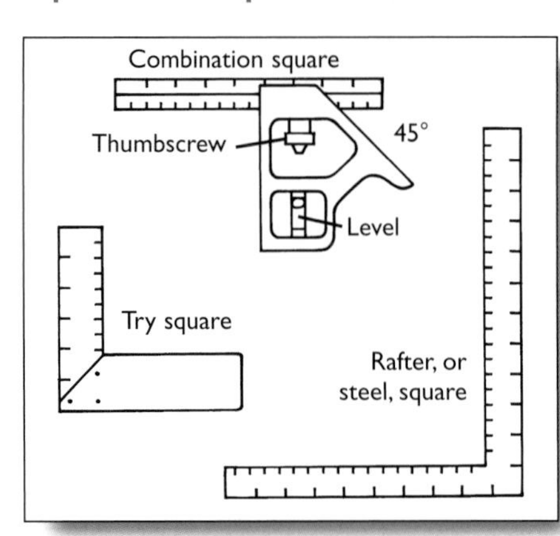

A *combination square* is useful for testing squareness and marking perpendicular, parallel or 45 degree lines. The versatility of the combination square makes it a good first choice.

The *framing square* is handy for lining up materials evenly, measuring and marking square cut lines. It is usually made of metal.

The *try square* is smaller than the framing square. It is also used for lining up, marking and to check cuts for square. One side is usually made of wood or metal.

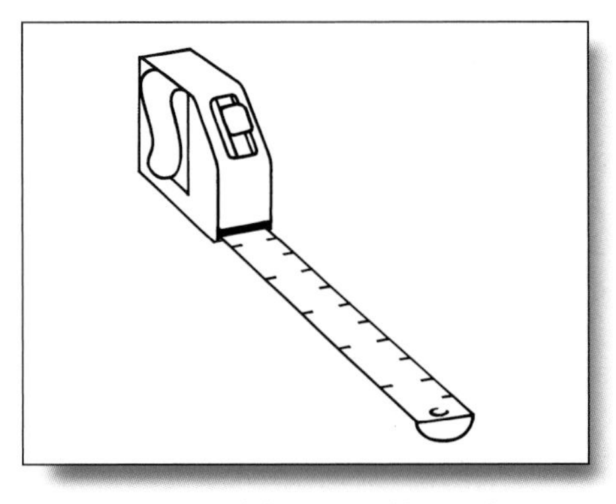

A 5-m (16 ft.) *retractable steel measuring tape* is an essential tool.

Staple guns

A staple gun and the appropriate staples are handy for a number of household chores such as replacing screens in wood windows and doors, tacking down carpet and other jobs that would otherwise call for small nail and a hammer. When using a staple gun, press down against the top of the gun with one hand as you squeeze the trigger with your other hand. Never discharge the gun in anyone's direction. The staples can cause serious eye injuries.

Trowels and jointers

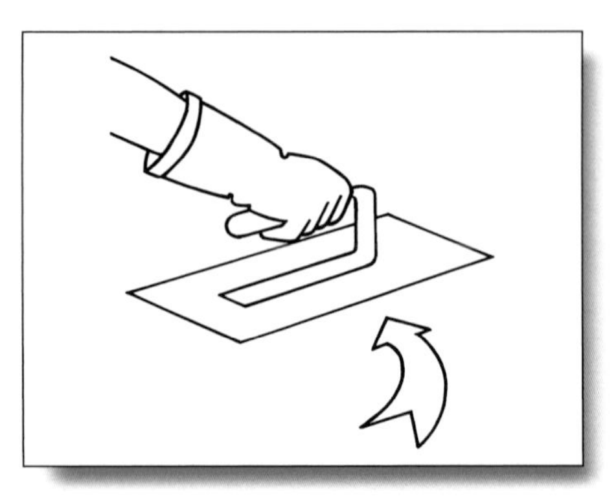

A masonry trowel has a flat, thin steel blade set into a handle. Use a *rectangular steel trowel* to spread concrete or apply cement parging.

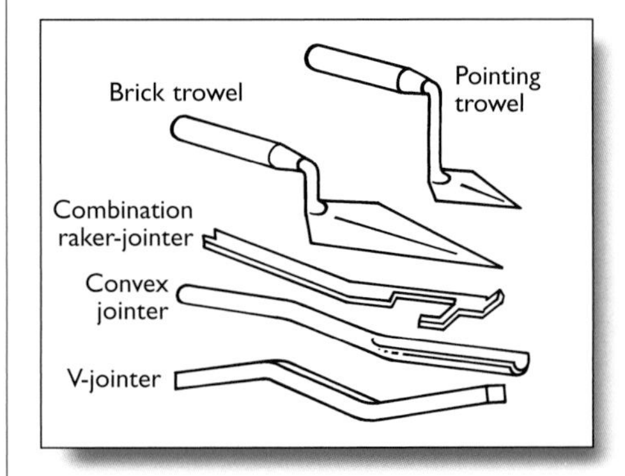

The *brick trowel* is a large triangular trowel and is used for mixing, placing and spreading mortar on bricks and blocks. Use a smaller, triangular, pointing trowel to fill holes and repair mortar joints in a process called pointing.

Use a *jointer* to finish masonry joints. Finish joints are made on the outside of a masonry wall to make it more waterproof and to improve its appearance. Jointers are available in different shapes.

Wrenches

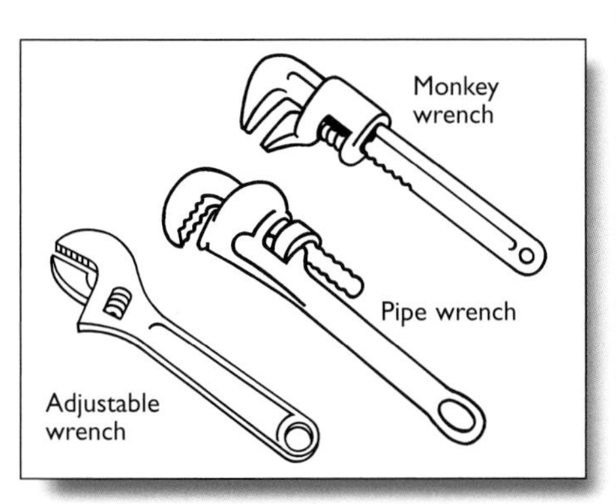

Adjustable wrenches are very versatile because they can be adjusted to fit different sizes of nuts or pipes. Use a *pipe wrench* to tighten or loosen pipes. Use a *monkey wrench* on large nuts.

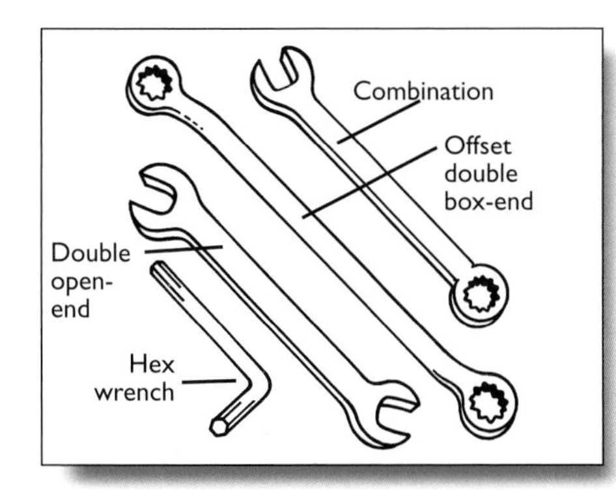

Fixed wrenches are available in open-ended, box-ended and hex styles. A popular choice is the combination wrench, which has an open end and a box end of the same size. A fixed wrench fits only one nut size, so you need an assortment of them. If a nut is hard to loosen, put a few drops of penetrating oil or kerosene on it. Let it soak a couple of hours or overnight.

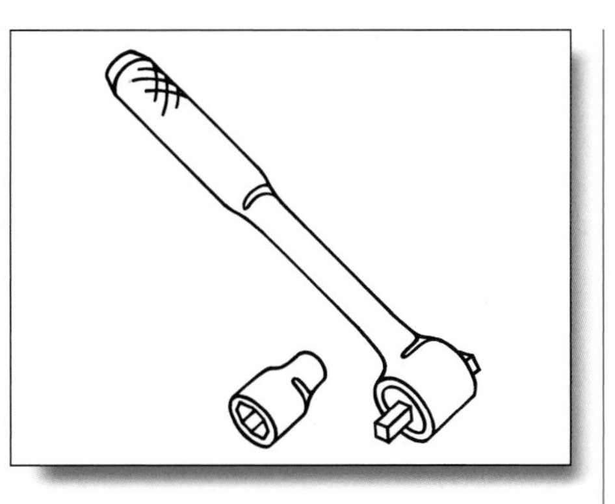

A *ratchet handle and sockets* are used to tighten or loosen nuts. They are stronger and faster than adjustables and usually provide more leverage than fixed wrenches. With the use of extensions, they work well for nuts that can't be reached with a wrench.

Other handy items

- Use an *all-purpose penetrating lubricant* to loosen rusted or "frozen" nuts.
- Have a selection of *abrasives* such as *sandpaper* and steel wool for sanding and finishing work.
- Use a *chalk line* to mark a straight line.
- *Clamps* are very versatile for holding objects while you work. They come in a variety of shapes, sizes and types for specialty uses.
- A *drain auger*, sometimes called a plumber's snake, may be needed when a plunger fails to clear a clogged drain.
- A *drop cloth* made out of fabric will protect items and not be as slippery as a plastic cloth.
- Keep a *flashlight* with charged batteries with your tool kit so it will be there when you need it.
- *Kneepads* are essential for floor or roof work where kneeling is required.
- A *stud finder* will help you to locate studs quickly and accurately.
- *Utility knives* can be used to cut almost anything; they come with replaceable blades that are razor sharp. The better knives have retractable blades.

Basic Home Repairs and Maintenance

Basic Home Repairs and Maintenance

FOUNDATIONS AND BASEMENTS

The most common foundation and basement problems are:

- minor cracks
- excessive mold growth

Maintenance includes:

- reducing moisture for all types of foundations and basements
- repairing above-grade parging

Prevention tips

Keeping water away from the foundation is the best way to prevent problems.

- Keep eavestroughs, downspouts and downspout extensions clean, in good repair, and extended far enough away so that the roof water flows away from the foundation.
- Add eavestroughs, if not present on all eaves.
- Ensure that the grade adjacent to the foundation does not "hold" water, but directs the surface water away.
- In the basement, check for moisture by taping a 1 m^2 (3 ft^2.) sheet of polyethylene to the concrete or masonry wall or slab. If condensation forms under the sheet after a day, there is moisture present.
- If the grade cannot be built up around the foundation, consider covering the ground with a layer of clay. Another option is to bury polyethylene sheeting 150 mm (6 in.) deep, and sloped away from the foundation. Either these methods will help to move water away from the foundation.
- Place screens on eavestroughs or downspout entries to prevent clogging.
- Install a sealed cover over the sump pump pit.

Special considerations

You may not see liquid water entering through cracks. However, water from the ground may be wicking into the concrete floor or walls and evaporating into the basement air. White, chalky stains, known as efflorescence, are an indication of moisture evaporation from concrete. The added moisture load in the air may lead to condensation and mold problems on cold surfaces.

Healthy Housing™

- The basement is the most likely place to find moisture and mold problems. Maintaining a clean and dry basement provides the foundation for a healthy home. Eliminate stored items that can hold moisture and become moldy. Use plastic containers, not cardboard boxes, and keep stored items on shelves, not on the floor.

Safety

- Most foundation problems relate to either water vapour being transferred from the ground into the house through walls and floor or liquid water entering through cracks. Minor, hairline cracks require attention to prevent water or soil gas entry. Major cracks can allow large quantities of water or soil gas to enter. Severe or active cracks (particularly if they're horizontal) may be an indication of unsafe conditions or future problems that could lead to collapse. Consult a structural engineer or basement specialist concerning multiple, severe or expanding cracks.

Tasks

Repair minor cracks

Minor cracks may not allow water leakage or indicate structural problems. However, they may provide a path for water vapour or soil gases to enter the home and should be sealed.

Skill level rating: 2 - Handy homeowner

Materials: polyurethane caulking, hydraulic cement

Tools: Hammer, cold chisel, wire brush, vacuum, caulking gun

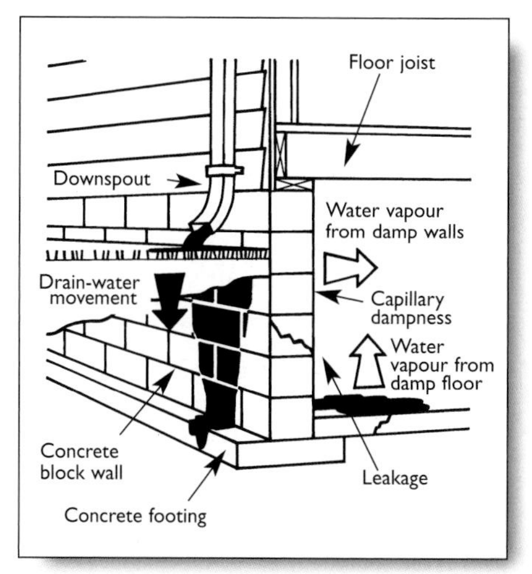

1. Look for cracks in the basement walls outside, above grade.
2. Inside the basement, look for cracks in concrete floors and walls or in concrete block walls. There are often minor cracks in the floor along the perimeter of the walls or around posts.
3. For narrow, dry cracks brush out any loose concrete chips or dust and vacuum the crack. Apply polyurethane caulking.
4. For slightly larger or wetter cracks, enlarge the crack with the hammer and cold chisel. Try to make the crack wider inside than at the surface to lock the patch in place. Brush and vacuum out the concrete chips and dust. Apply a hydraulic cement patch following the manufacturer's instructions.

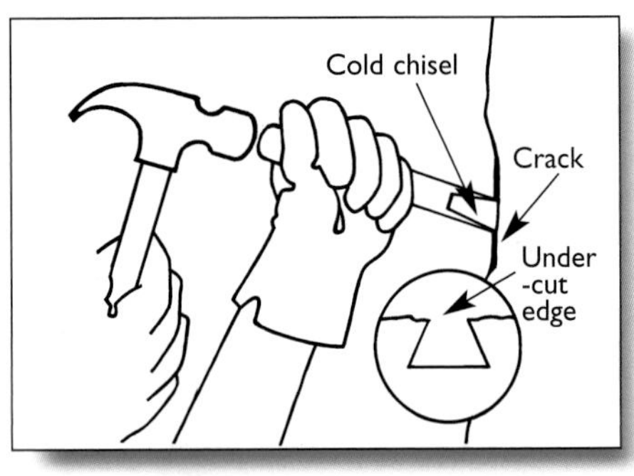

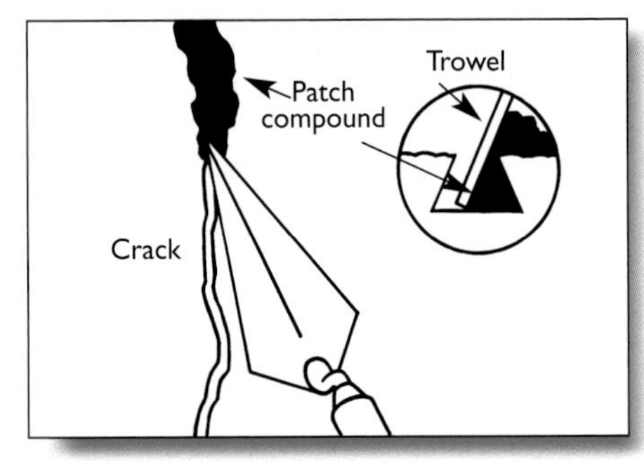

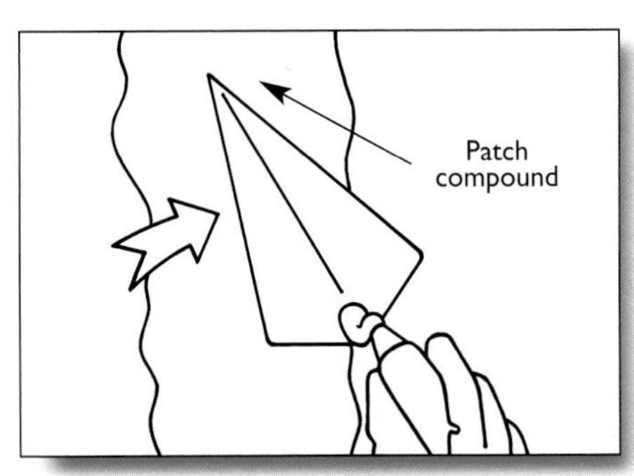

5. Note: Usually if there is an interior vertical crack on a concrete foundation it will continue all the way to the exterior. If there is moisture seeping through, the exterior may have to be excavated to expose the foundation to the footing. Chisel out the crack according to the hydraulic cement manufacturer's instructions, usually about 19 mm (3/4 in.) wide and 25 mm (1 in.) deep so that the edges are undercut. Install hydraulic cement following the manufacturer's instructions, then apply foundation coating. For extra protection, place a 300 mm (12 in.) wide polyethylene or bituminous membrane over the crack to grade level, sealing as required with foundation coating or according to manufacturer's instructions.

Install a moisture barrier in a dirt floor basement or crawl space

Whether dirt floor basements or crawl spaces are heated or not, they may create excessively high water vapour levels in the air. With just a dirt floor, there is no barrier to prevent the entry of water vapour or soil gases. Traditionally, crawl space vents have been used to keep crawl spaces dry but research has shown that these vents are often ineffective. Also, in spring and summer when the outside air is warmer and carries more water vapour in suspension, the vents may actually introduce more moisture to the crawl space. However, an uncovered dirt floor can be the biggest moisture source. Even if it seems dry, it acts as a giant sponge, wicking water to the surface where it evaporates into the crawl space air and possibly condenses again on cooler surfaces. Very few crawl spaces are actually isolated from the house—so the crawl space air becomes the house air eventually. A ground moisture barrier is needed to control moisture in crawl spaces.

Uncontrolled moisture may lead to mold growth. Mold growth may occur even on the dirt under a moisture barrier. Although moisture may not penetrate the barrier, toxic chemical emissions from mold may come through. It's best to create an inhospitable habitat for molds under the dirt floor moisture barrier, in the most environmentally friendly way possible.

Skill level rating: 2 - Handy homeowner

Materials: sidewalk salt, 0.15 mm (6 mil) or preferably thicker polyethylene sheeting, contractor's sheathing tape, polyurethane caulking; possibly concrete, patio stones, concrete blocks or bricks

Tools: shovel, garden rake, caulking gun, (a tape dispenser is handy), staple gun, garbage bags, dust mask, goggles, old clothes or disposable coveralls

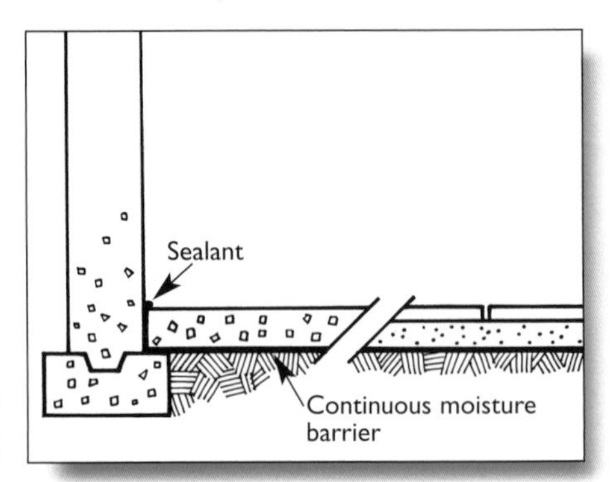

1. Ensure that the grading and drainage around the house will prevent crawl space flooding. These issues must be resolved first. Installation of a floor drain would also be ideal but may be impractical.
2. Always wear a mask and goggles when working in the crawl space. If there is standing water or visible mold, seek professional assistance.
3. Place any debris from the crawl space into garbage bags and carry it out.
4. Rake the crawl space dirt floor smooth. Remove any sharp rocks. Ideally, grade the dirt floor so there are no depressions likely to become pools of water.
5. Sprinkle sidewalk salt throughout to create an inhospitable habitat for mold growth.
6. Lay the polyethylene moisture barrier on the floor. Overlap and tape the joints. Caulk the edges to the base of the wall or, if possible, staple and seal to the main floor sill plate at the top of the wall.
7. If possible, install a poured concrete floor. Next best is to cover the polyethylene with a layer of sand and install a floor made of patio stones. If neither option is possible, try to ensure that the moisture barrier is undisturbed by holding it in place with concrete blocks or bricks.

Prevent mold from forming in foundations and basements

The secret to preventing basement mold problems is to keep basements dry. Mold cannot grow without moisture. Basement moisture and mold symptoms include foundation cracks that leak, flooding from a high water table, damp or moldy walls or floor, condensation on windows or pipes, wet insulation, moisture damaged finishes, musty or damp carpets, damp stored items, high humidity, and stuffy damp smells. Some basic maintenance techniques can keep moisture to a minimum.

Skill level rating: 1 - Simple maintenance

Materials: none

Tools: garbage bags, utility knife, dust mask, goggles

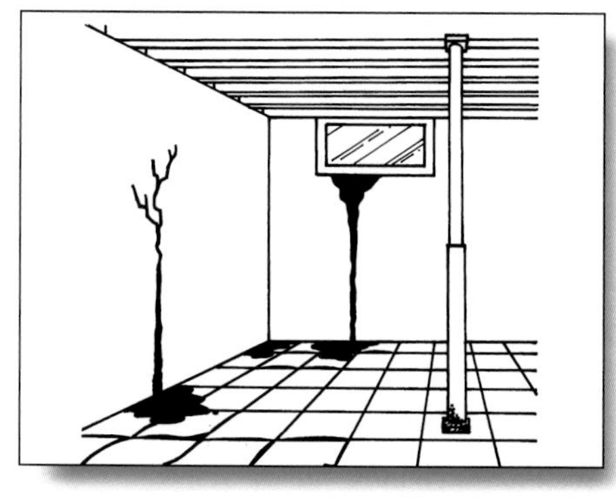

1. Ensure that eavestroughs and downspouts (with extensions) direct water away from the building.
2. Slope the grade away from the house.
3. Where a perimeter drain tile system exists, remove leaves or debris from window wells so that they can drain more easily.
4. Patch cracks (as noted above).
5. Remove basement carpets unless the concrete floor is absolutely dry. Carpets trap moisture, providing a good habitat for dust mites and mold. Even carpets that seem dry may be damp underneath. If necessary, cut carpets into strips and place in garbage bags for removal.
6. Minimize basement storage to provide less food for molds and improve air circulation. Store items on shelves off the floor and away from the exterior walls. Keep items on the shelves in plastic containers, not in cardboard boxes. Cardboard boxes absorb moisture and stay damp.
7. If there is ever a flooding event, remove all damaged materials promptly.
8. Use a dehumidifier in the basement in summer. Circulating air with a fan can be helpful in all seasons.

Repair above-grade parging

Parging provides a finished appearance above grade, but it also serves as part of the moisture control system for foundations both above and below grade. Above grade, parging helps prevent rain or surface water leakage by sealing any cracks, holes or form tie cavities (created by the ties that hold the concrete forms together). It also protects exterior insulating systems that may be damaged by exposure to sunlight. Below grade, parging must be applied to concrete block walls to seal joints and provide a smooth surface similar to poured concrete, before dampproofing is applied. Parging can also be used to seal form tie holes or other openings in poured concrete walls before dampproofing. Being struck with a hard object may damage specialty parging systems that cover exterior insulating systems. Parging on galvanized stucco mesh over dense glass fibre, mineral fibre or preserved wood foundations may also suffer damage. Parging over concrete or block may flake off due to the freeze-thaw action of water that may get behind it.

Skill level rating: 2 - Handy homeowner

Materials: Portland cement, sand and water; or a specialty pre-mixed parging designed to adhere to polystyrene

Tools: hammer, cold chisel, stiff brush, trowel, sponge, shovel, wheelbarrow or mixing board, pail

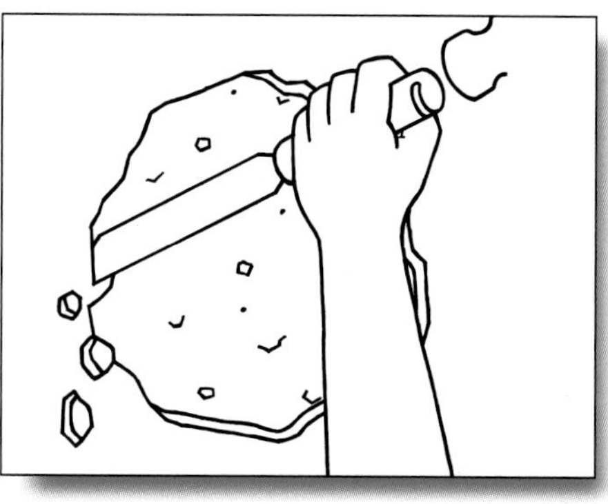

1. Remove any loose parging with the cold chisel. Clean the affected area with the brush.
2. Dampen the area to be patched.
3. Combine three parts screened sand to one part cement (3:1 mix) with enough water to make a stiff mix.

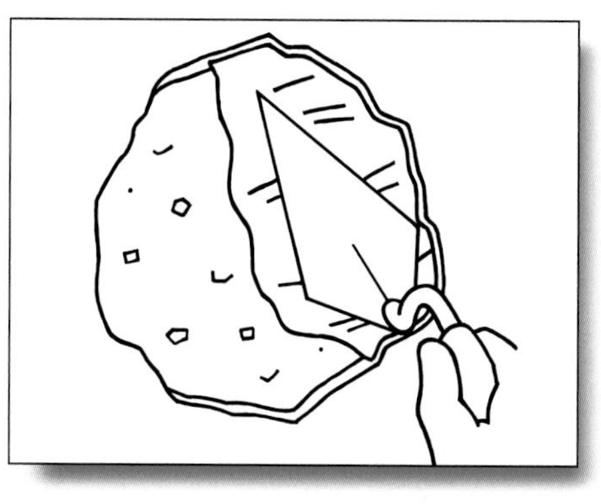

4. Trowel a layer no more than 6 mm (1/4") thick onto the affected area.

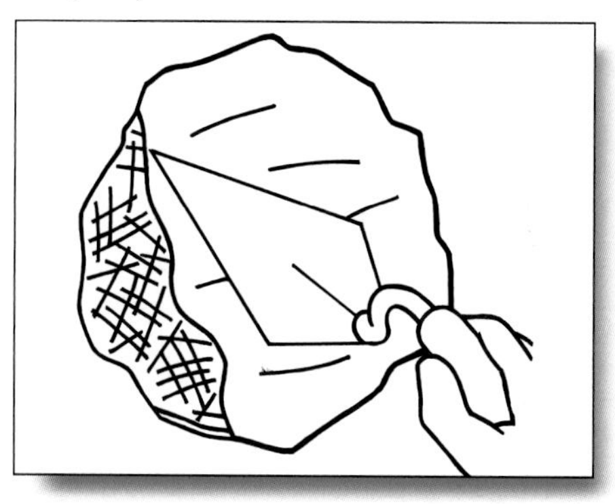

5. If a second layer is required, when the patch is still soft, scratch it with a nail or other sharp object to provide a gripping surface for the next coat.

6. When the patch has dried thoroughly (24 hours), dampen the area again and apply a second 6 mm (1/4") layer that should blend smoothly with the surrounding parging.
7. When the patch is close to dry, texture it to match the surrounding parging using a damp sponge.

Clean up visible mold

The presence of mold is a sign that there is too much moisture in your home —a situation that must be corrected.

Mold may be any colour: black, white, red, orange, yellow or green. If you're not sure whether the suspected spot is mold, dab it with a drop of household bleach. If the stain loses its colour or disappears, it is likely to be mold. If there is no change, it is probably not mold.

The amount of mold determines the type of clean-up needed.

If there are one or more patches of mold, at least one of which is greater than 3 m² (32 ft²), about the size of a full sheet of plywood, it is considered "large" and should be cleaned up by a professional mold cleanup contractor. A mold problem this extensive also indicates a major moisture problem that must be addressed to prevent the mold from returning.

"Small area" clean-up

You can clean up "**small areas**" of mold (fewer than three patches, each smaller than a square metre) yourself. Infants and other family members with asthma, allergies or other health problems should not be in the work area or adjacent room during the cleaning.

Skill level rating: 2 - Handy homeowner—small area of mold

Materials: unscented detergent solution, wet rag or sponge, baking soda

Tools: household rubber gloves, disposable dust mask, safety glasses or goggles, high efficiency particulate air (HEPA) vacuum

Washable surfaces

1. Vacuum surfaces with either a vacuum cleaner equipped with a high efficiency particulate air (HEPA) filter, an industrial vacuum cleaner exhausted outside or a central vacuum exhausted outside.
2. Scrub with an unscented detergent solution.
3. Sponge with a clean, wet rag and dry quickly.

Using an unscented detergent will make it easier for you to detect residual moldy odours.

Moldy drywall

1. If the mold is only on the surface and not caused by a moisture problem behind the drywall, follow the vacuuming instructions listed above, then clean the surface with a damp rag using baking soda or a bit of detergent. Do not allow the drywall to get too wet.

Mold that comes back after cleaning usually indicates that the source of moisture causing the mold growth has not been removed. Seek professional help from a trained indoor air quality investigator.

How to clean up moderate mold problems

If you follow the proper procedures and use the proper protective equipment, you can clean up "**moderate areas**" of mold. "***Moderate***" means more than three patches of mold, each smaller than one square metre, or one or more isolated patches larger than one square metre but smaller than 3 m² (32 ft²) or about the size of a standard sheet of plywood.

Infants and other family members with asthma, allergies or other health problems should not be in the work area or adjacent room during the cleaning.

A small cleanup should take minutes (not hours) to finish. When the cleanup will take hours to a day to finish, it is suggested that you upgrade to a better safety mask, such as a half- or full-face respirator with HEPA cartridges. These masks need to be properly fitted to your face.

Skill level rating: 3 - Skilled homeowner

Materials: plastic sheeting, tape, unscented detergent solution, wet rag or sponge, baking soda, TSP solution for concrete surfaces

Tools: household rubber gloves, disposable dust mask or half-face respirator with HEPA cartridges, safety glasses or goggles, exhaust fan, HEPA vacuum

General cleaning

1. Isolate the area to be cleaned with plastic sheeting, taped to walls and ceiling.
2. Install an exhaust fan on a window in the room being cleaned to prevent contamination of other areas of the house as well as to provide ventilation.
3. Vacuum surfaces with a vacuum cleaner that has a HEPA filter or a central vacuum that exhausts directly to the outside.
4. Scrub or brush the moldy area with a mild, unscented detergent solution.
5. Rinse by sponging with a clean, wet rag. Repeat. Dry quickly.
6. HEPA vacuum the surfaces that were cleaned as well as surrounding areas.

Cleaning wood surfaces

1. Vacuum mold from wood surfaces using a HEPA or externally exhausted vacuum. Skip the vacuuming step if the wood is wet.
2. Clean with a detergent solution, then sponge with a clean, wet rag.
3. Extract the moisture using a dry/wet vacuum and/or clean, dry rags.
4. Accelerate the drying with fans and open windows. If the relative humidity outside is high, close windows and use a dehumidifier. The wood should not be allowed to remain wet for more than a day.

If cleaning with detergent and water does not remove the mold, wear a mask and try sanding the surface with a vacuum sander (simultaneous vacuuming and sanding). Do not attempt to sand without vacuuming, because it will spread the mold. This method will not work if the mold has penetrated to the core of the wood. Severely moldy and rotten wood should be replaced.

Cleaning concrete surfaces

1. Vacuum the concrete surfaces to be cleaned with a HEPA or externally exhausted vacuum cleaner.
2. Clean up surfaces with detergent and water. If the surfaces are still visibly moldy, use TSP (trisodium phosphate). Dissolve 250 ml (8 oz.) of TSP in 9 L (2 gal.) of warm water. Stir for two minutes. Note: **TSP must not be allowed to come in contact with skin or eyes. Wear rubber gloves and eye protection.** Saturate the moldy concrete surface with the TSP solution using a sponge or rag.
3. Keep the surface wet for at least 15 minutes.
4. Rinse the concrete surface twice with clean water.
5. Dry thoroughly, as quickly as possible.

Moldy drywall

The paper facings of gypsum wallboard (drywall) grow mold when they get wet and don't dry quickly. Cleaning with water containing detergent not only adds moisture to the paper but also can eventually damage the facing. If the mold is located only on top of the painted surface, remove it by general cleaning (above). If the mold is underneath the paint, the moldy patch and other moldy material behind it are best removed by cutting out the section of moldy drywall. The surrounding areas should also be cleaned. If the area is large, a mold clean-up contractor should do this. New materials will become moldy if the moisture entry has not been stopped. Materials should not be replaced until the source of the moisture is corrected. Before removing moldy wallboard, temporarily cover the affected areas with plastic sheeting sealed at the edges with tape. This helps to minimize the spread of mold spores during the remediation.

Any areas that show new patches of mold should be cleaned promptly.

Call a trained investigator

A trained investigator is needed when:

1. There is an extensive amount of mold.
2. Mold comes back after cleaning.
3. The house smells musty but no mold is visible.
4. A family member is sick and the house is suspected of having a mold problem.

Contact your local CMHC office for a list of trained investigators in your area.

FLOORS

Floors take a lot of wear and tear. We walk on them, drag things across them and drop things on them. Damaged flooring should be repaired immediately to prevent accidents. Keeping floors in good condition requires general cleaning and regular maintenance.

The most common floor problems are:

- damaged or loose sheet floor covering
- damaged flexible floor tiles
- loose or damaged ceramic floor tiles
- buckled, stained or torn carpeting
- scratched or worn hardwood floors
- squeaking

Maintenance includes:

- regular sweeping and washing to minimize dirt that will scuff and scratch the floor.

Prevention tips

- Fix small problems before they turn into big repairs.
- Have mats available outside your entry door to encourage people to wipe their feet and minimize the amount of tracked-in dirt.
- Have people remove their shoes in the house to avoid tracking dirt through the house.

Repair tips

- When installing flooring, keep some extra pieces in case you need replacement parts for future repairs.
- If you install new vinyl floor covering, try to choose a heavyweight type, especially for high traffic areas. The flooring will wear better and need less maintenance.
- A putty knife is handy for removing flooring and applying glue.
- Use a rolling pin to smooth the patched area.
- The largest tile selection will typically be found at a store that specializes in tiles and flooring products.
- A special carpet-cutting tool, similar to a cookie cutter, is available from carpet stores.
- It's a lot easier to cut ceramic tiles with a proper tile cutter. Tile cutters are often available at tool rental stores.
- Always sand wood floors in the direction of the grain, not across the grain.
- Wear rubber gloves to protect your hands from stain.
- Specialty tools such as manual or electric tile cutters, nippers and trowels can be rented.

Special considerations

Healthy Housing™

- Choose water-based glues for flooring repairs. These products have minimal affects on the indoor air quality in your home, can be cleaned up with water and cause few disposal problems.

Safety

- Older flexible tile flooring (usually before the mid-1980s) may contain asbestos. Asbestos poses health risks only when fibres are in the air for people to breathe. For a small repair such as replacing a broken asbestos floor tile, wet the material to minimize dust, wear an approved dust mask and protective clothing, and do a thorough cleanup using a vacuum with a HEPA filter. Check with your local and provincial authorities about how to safely dispose of the asbestos material in your area.
- Large repairs such as replacing an entire floor that is covered with asbestos tiles, will have to be done by a contractor who is experienced in asbestos removal. Anybody who works with asbestos must wear an approved face mask and gloves, along with protective clothing. Additional precautions such as isolating the workspace, filtering the exhaust air and disposing waste properly are usually required.
- Damaged floors can pose serious tripping hazards.

Tasks

Fix vinyl floor covering

Skill level rating: 2 - Handy homeowner

Materials: replacement flooring, masking tape, flooring glue

Tools: straightedge or square, utility knife, putty knife, rolling pin

1. Cut a piece of flooring large enough to cover the area to be patched and to match any pattern. Tape the replacement piece in place, lining up any patterns with the existing flooring.

2. Using the utility knife, cut through both pieces of flooring to create a perfect patch.

3. Remove the entire patch piece and the damaged piece of original flooring.

4. Apply glue to the patch piece and carefully press it into place in the hole. Finish by flattening the patch with a rolling pin, then wiping off any extra glue.

Replace flexible floor tiles

Skill level rating: 2 - Handy homeowner

Materials: replacement tiles, flooring glue if tiles are not self-stick

Tools: straightedge or square, utility knife, heat gun

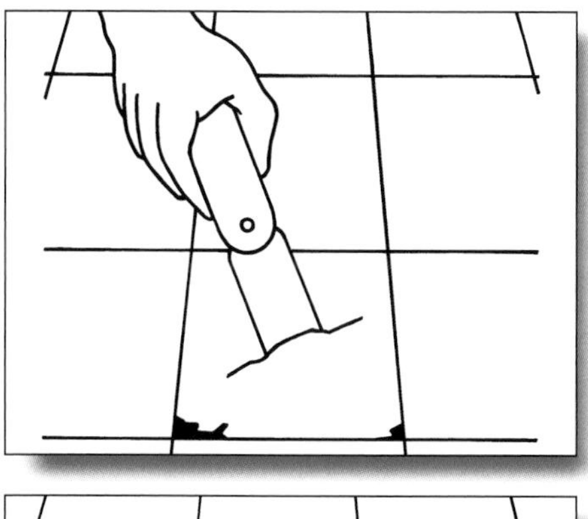

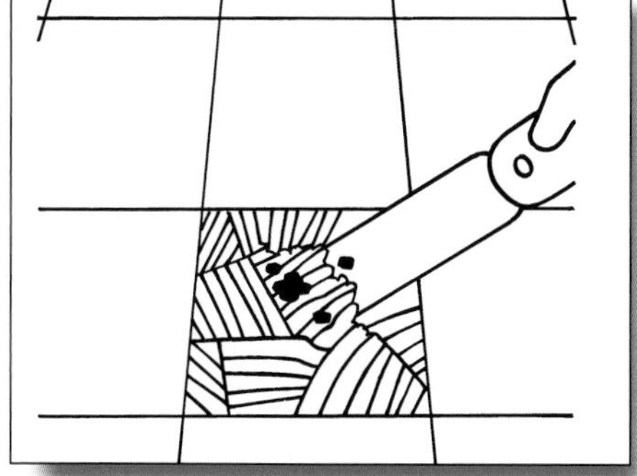

1. Lift the old tile from the subfloor with the help of the putty knife. A heat gun will help to soften the adhesive and make it easier to remove the tile. Clean off the adhesive from the subfloor. The subfloor and replacement tiles must be clean of adhesive and dirt in order for the repair to work.
2. The tiles and subfloor should be kept at room temperature or warmer before you attempt repairs. Colder temperatures make the tile stiff, more likely to break and prevent the adhesive from adhering properly.
3. Align the replacement tile with the pattern on the existing floor. Check to be sure that it fits in the old opening before applying adhesive or peeling off the protective backing from self-adhesive tiles. If it does not fit or the tiles overlap, trim it with the utility knife.

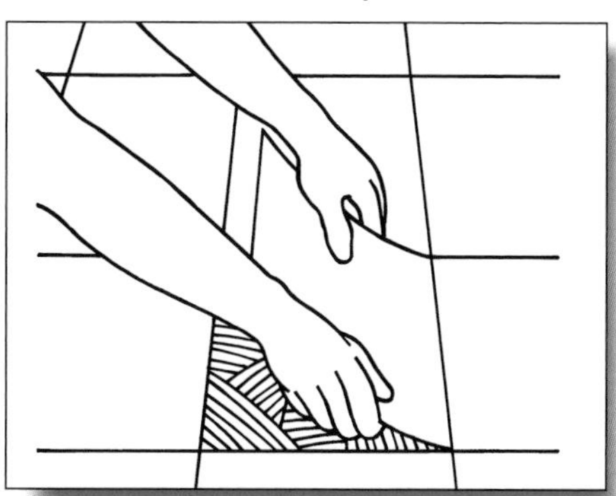

4. Check the trimmed tiles for fit. If all is okay and the subfloor is still clean, you are ready to begin resetting. Always start with full tiles and work out toward the trimmed ones. If the new tiles are self-adhesive, just peel off the protective coating, line up the tile carefully, and firm it in place. To set non-adhesive tiles, apply a thin, uniform coat of tile adhesive to the subfloor with the applicator or putty knife before firming each tile in place. Use a rolling pin to firm in place.
5. Clean up excess adhesive immediately according to the manufacturer's directions.

Replace ceramic floor tiles

Skill level rating: 2 - Handy homeowner

Materials: grout, tile adhesive or mortar, tile

Tools: pencil or non-permanent marker, mixing bowl, utility knife or glass cutter, combination square or straightedge, pliers, measuring tape, scrap piece of wood to trim tiles on, notched tile adhesive or mortar applicator, nippers, trowels, tile cutter, rubber gloves, protective eyewear

1. Remove grout around the tile with a utility knife. Lift out the tile. If the tile is hard to remove, break the tile into small pieces with a hammer. Wear protective eyewear to protect your eyes from flying tile. Scrape the old adhesive and other loose material off the floor and any existing tiles if you are going to use them again.

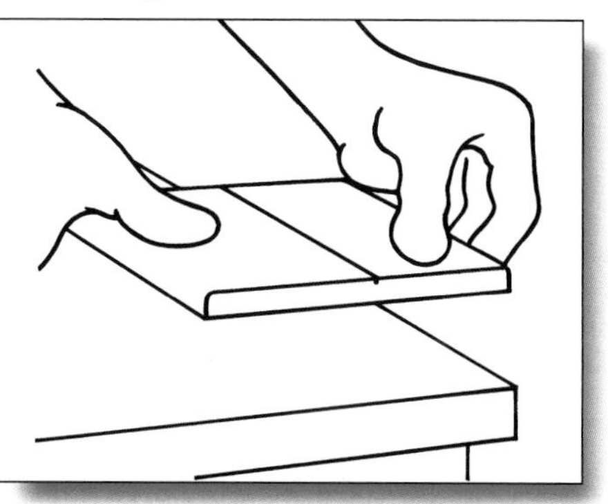

2. If you are using a new tile, you may have to cut it to fit. Mark the cut line or lines on the top (glossy) surface of the tile using a pencil or non-permanent marker. Make a straight cut by lining up a straightedge or square with the marked lines and scoring (shallow surface cut) the surface with a sharp utility knife or glass cutter. You must press down heavily enough to cut through the tile glaze, so keep your fingers clear of the blade. Afterwards, place the scored line over the edge of a table top, press the tile down firmly with the palm of one hand, and snap the overlapping edge off with the other hand. The tile should break cleanly off at the line. (Renting and using a proper tile cutter makes this much easier to do.)

3. Cut curved or round lines as a series of short interlocking straight scores made just inside the area perimeter. Score the enclosed area with hatches, then carefully nip out the enclosed, hatched area with pliers. Rough edges can be ground to shape with an inexpensive grinding bit mounted on a power drill.
4. Test fit the tile, then spread ceramic tile adhesive on the tile back and the space to be filled. If using mortar, spread the mortar evenly on the space to be filled using the notched trowel. Press the tile or tiles firmly in place, in line with the existing tile edges. Clean out excess adhesive from grout spaces.
5. Once the tile has set firmly, fill the joints with grout. Mix only as much as you need, adding water to the powder to form a paste. Press the mixture into the joints with a putty knife and finish tooling the joints with a wet, gloved finger.

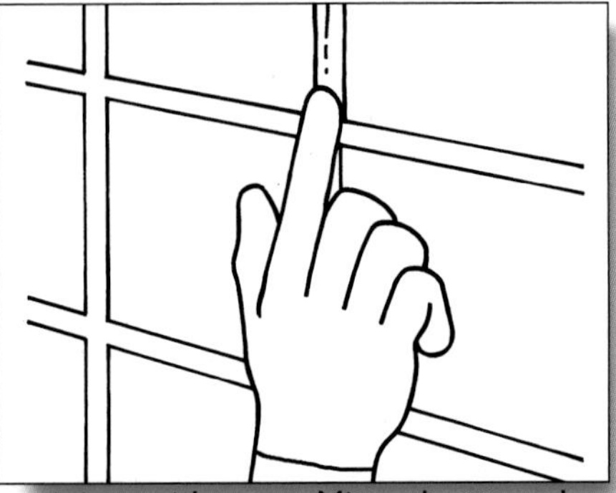

6. Wipe the excess grout from tiles and other surfaces. Sponge grout or mortar off tools in a shallow pail of water. Wipe tools clean with a dry rag.
7. Let grout dry for about an hour. Do a finish polish with a soft, damp cloth.
8. Dispose of the leftover grout in your garbage, not down your sinks!
9. Let the newly grouted joints dry overnight.

Repair damaged carpeting

Skill level rating: 2 - Handy homeowner

Materials: replacement carpet, double-sided tape, seam adhesive

Tools: straightedge or square, utility knife or carpet tool

1. Cut away damaged carpet using the carpet tool or utility knife and straightedge.
2. Cut a replacement patch to fit the hole. Be sure to line up the pattern and nap of the patch with the carpet. Cut double-sided carpet tape slightly larger than the hole. Insert the tape into the hole so that it goes under the carpet and patch seam.

3. Line up and place the patch so that it matches the carpet pattern and direction of the nap. Seal the seam with the seam adhesive.

Repair hardwood floors

Skill level rating: 2 - Handy homeowner

Materials: wood filler and wood stain to match

Tools: putty knife, fine sandpaper, cloth, rubber gloves

1. Clean dirt and debris from the area to be repaired. Patch the hole or scratch with wood filler, using a putty knife.
2. Gently sand the patch until it is smooth and even with the surrounding area.
3. Apply stain to the patched area so it matches the rest of the flooring.
4. Seal the patched area with the same coating as the floor.

Stop floor squeaks

The structure of a house floor consists of wood joists that are supported by load-bearing walls, wood beams or steel beams. Solid blocking or cross-bridging may be used between joists so loads can be shared. On top of the joists is a subfloor made of plywood, oriented strand board or, in older homes, wood boards. If the floor finish is sheet vinyl, vinyl tile or ceramic tile, another layer of plywood or similar product may be installed on the subfloor to provide a smooth surface under those materials. Hardwood and carpeting may be installed directly over the subfloor.

Floors squeak for a variety of reasons. If the pieces of wood in the layers are not tightly fixed to each other, they may have enough flexibility to rub against each other or against the fasteners. Pieces may not have been adequately fitted or fastened originally. Solid wood pieces may have dried and shrunk so fasteners are no longer tight. Subfloor panels may squeak at the edges where they meet. Panels may de-laminate (glued layers separate) due to moisture. In older homes, the floor structure may not be rigid enough to prevent squeaky floors. In order to stop squeaks, all pieces must be properly supported and tightly fastened.

The best time to fix floor squeaks is before you install a new floor finish. The first challenge is to pinpoint the squeak. This is usually easiest if the joists and subfloor are visible from below. In the squeaky area, examine the joists and subfloor while a helper walks on the floor area above. If there is a finished ceiling below the squeaky floor, you'll have to try to locate the squeak from above. You'll also have to try to locate the joists by tapping the floor and listening for a solid sound that indicates a joist. Joists are usually found every 400 mm (16 in.).

Skill level rating: 3 - Skilled homeowner

Materials: annular-ringed finishing nails, common spiral nails, screws, glue, wood shims, wood glue, putty or wax, solid blocking, steel bridging,

Tools: hammer, screwdriver, nail set, handsaw or circular saw, drill, chalk line

From below

If the whole floor seems too flexible, rows of bridging or blocking not more than 2.1 m (6 ft. 10 in.) apart may stiffen it by sharing loads between joists.

1. Measure the length of the joists between supporting beams or walls underneath.
2. Divide the length so that one or two equally spaced rows of bridging will not be more than 2.1 m (6 ft. 10 in.) apart.
3. Chalk a line on the underside of the joists, perpendicular to their length, where bridging is to be located.

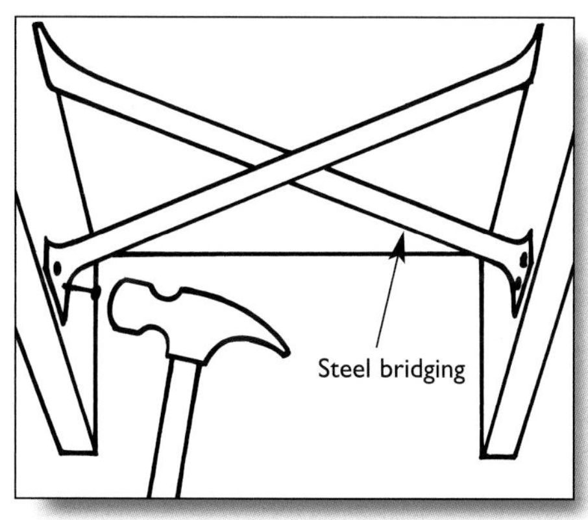

4. To install pre-manufactured steel bridging, hammer one end into the top of a joist and the other into the lower part of the adjacent joist. Set another piece at the opposite angle to form an X. Since there are wires and pipes between floor joists in most houses, using steel bridging is probably easiest. Wood bridging consisting of pieces not less than 19 x 64 mm (3/4 in. x 2 1/2 in.) or 38 mm x 38 mm (1 1/2 in. x 1 1/2 in.) can also be used.

Note: When metal bracing or heating ducts come into contact with copper plumbing pipes, an electrochemical reaction can occur that will cause the metals to corrode prematurely. To avoid this reaction, do not allow the two metals to come into contact. For example, ensure that heating ducts and copper plumbing pipes are separated and securely fastened. Use wood bridging instead of steel bridging where necessary.

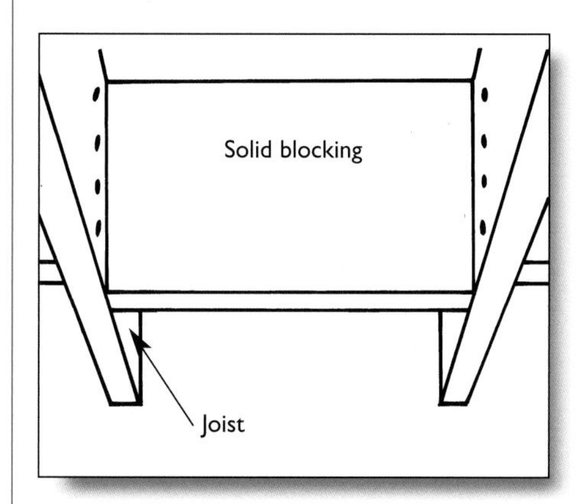

1. If you decide to use solid blocking, cut pieces from the same size lumber as the joists.
2. Fit the pieces in place in a staggered fashion to allow end nailing with 3 or 4 nails. Where necessary, use smaller dimension lumber to accommodate pipes or wires.

If there is a gap between a joist and the subfloor:

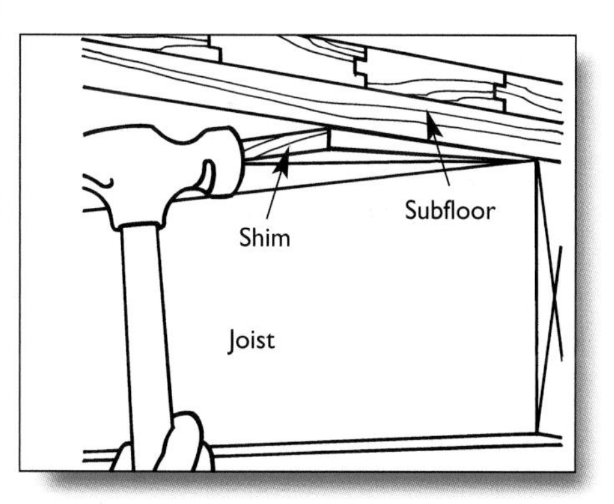

1. Lightly glue a shim and tap it into the gap between the joist and the subfloor. Be careful not to drive it in so hard that the subfloor rises.

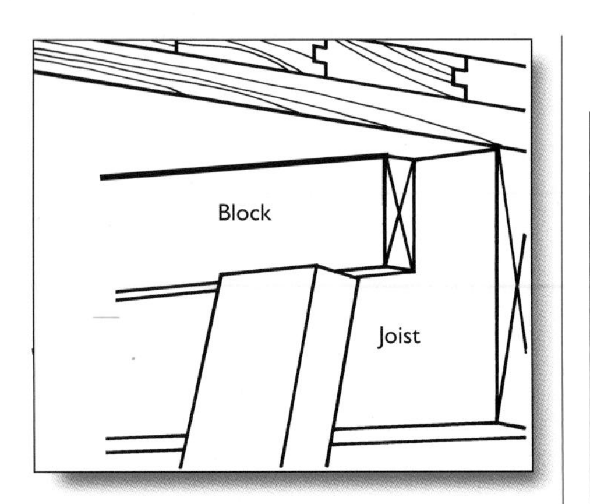

2. Alternatively, screw a lightly glued 19 mm (3/4 in.) or 38 mm (1 1/2 in.) strap or block to the side of the joist, tight to the underside of the subfloor.

If there seems to be a squeak between the subfloor and a buckled wood finished floor.

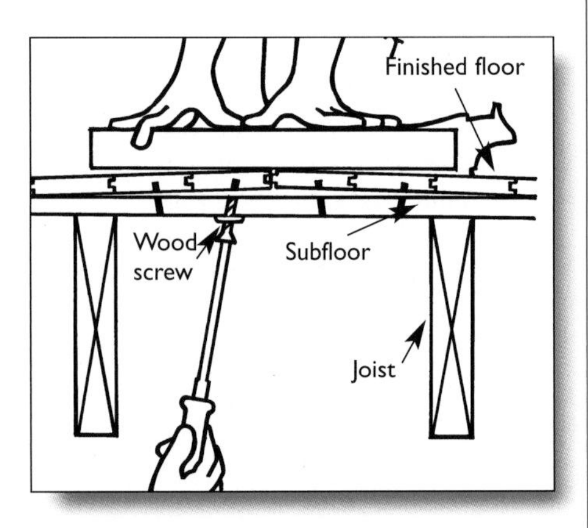

1. Hold the finished floor down with a weight (if possible).
2. Drill pilot holes through the subfloor and into the finished floor from underneath. Use a piece of tape on the drill bit as a depth indicator, to make sure that you don't drill through the finished floor.
3. Screw the subfloor and finished floor together using screws that are long enough to secure them both. Washers under the screw heads will prevent them from being drawn too far into the subfloor.

From above

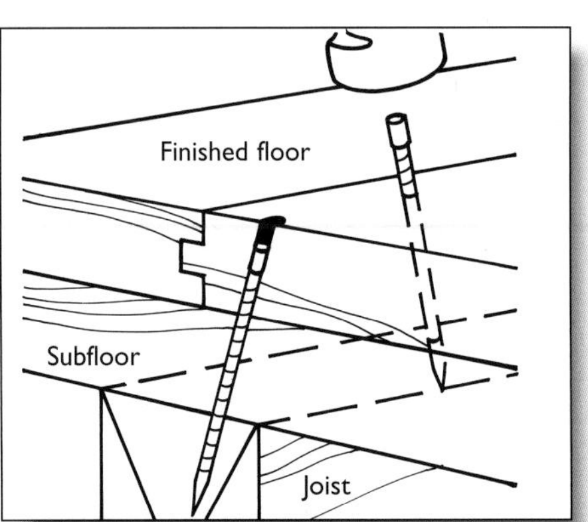

1. In hardwood floors, drill a suitably sized, angled pilot hole through the finished flooring over a joist.
2. Drive a finishing nail through the flooring into the joist.
3. Use a nail set to sink the nail below the floor surface. Use a suitable putty or wax to fill the hole.

It is very difficult to fix floor squeaks in non-wood flooring types from above without causing visible damage. It may be possible to drive and sink finishing nails through some carpets. The chance of success in stopping squeaks is limited. Removing the carpet first is recommended.

WALLS AND CEILINGS

Walls and ceilings make up the largest areas inside your home. Keeping them in good shape makes them easy to clean and helps to avoid bigger repairs.

The most common wall and ceiling problems are:

- holes and nail head pops
- cracks in the wall surface
- drywall taping pulling away from the surface
- cracks around the bathtub or shower
- loose or damaged ceramic tiles
- items such as handrails, curtain rods that have pulled away from the wall
- gaps around electrical boxes and plumbing breather vents that penetrate into attic

Maintenance includes:

- interior painting.

Prevention tips

- When preparing to paint, take the time to check and repair minor damage that could lead to bigger repair problems.

Repair tips

- Wallboard or drywall is easier for a non-professional to repair than plaster. If your home needs major plaster repairs, consider whether it's within your abilities and can be done within a reasonable period. Plaster repairs involve special techniques and can be messy. It's often worthwhile calling a professional.
- The largest selection of ceramic tiles will typically be found at a store that specializes in tiles and other flooring products.
- Specialty tools such as manual or electric tile cutters, nippers and trowels can be rented.
- Use care when estimating the amount of paint you will need for a job. Leftover paint will spoil so it can be stored only for short periods. Store cans upside down.

Special considerations

Healthy Housing™

- Use water-based paint for interior painting. These products have minimal affects on the indoor air quality in your home, can be cleaned up with water and cause few disposal problems.

Safety

- If your home is more than 40 years old, you should assume that the paint in your home contains lead. Lead-based paint is not dangerous if it is in good condition, but if it is peeling and flaking then the paint presents a potentially harmful situation. Sanding and scraping lead-based paint can also produce large amounts of dust that contains lead.
- Those especially at risk from lead-based paints are infants, young children, pregnant women and the fetus. Paint samples can be tested with a home test kit or through laboratory analysis. Current federal and provincial laws restrict the amount of lead that can be contained in commercial products.

Tasks

Patch small holes

Holes in drywall are usually caused by minor damage such as moving a picture or popped nails used to install the drywall. Popped nails occur because the framing lumber under the wallboard does not dry enough during installation. As the lumber dries, it shrinks causing the nails to pop through the wallboard, creating a hole.

Skill level rating: 2 - Handy homeowner

Materials: premixed drywall compound, drywall screws, paper or glass fibre drywall tape, paint

Tools: straightedge or square, drywall saw, glue gun, wallboard knives or scrapers, medium grit (80-100) sandpaper, small block of wood or sanding block, old cloth or paintbrush, utility knife, screwdriver

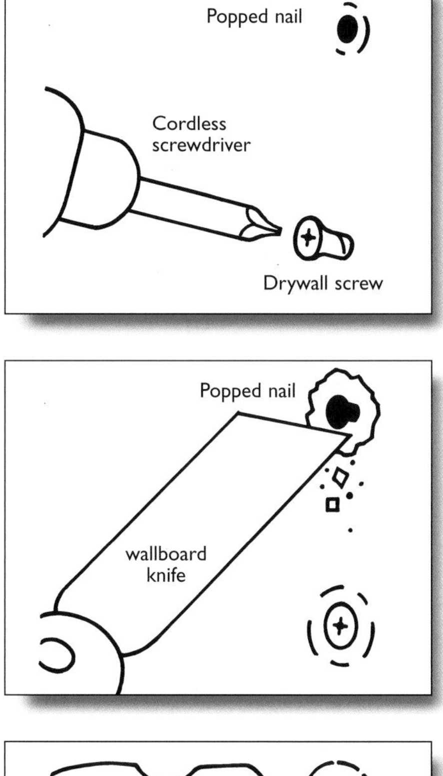

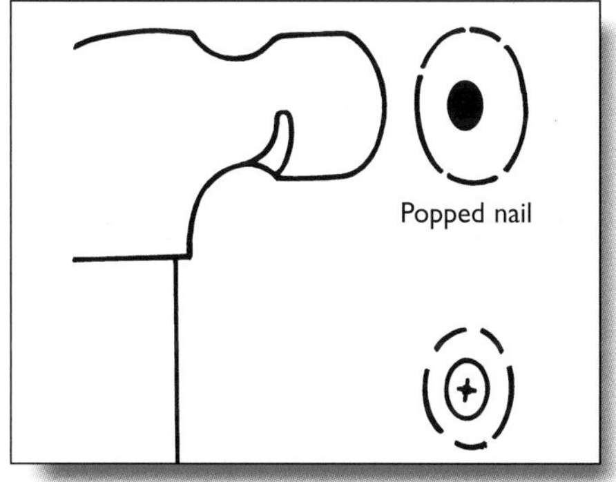

1. Press wallboard against the wall stud and fasten a screw about 50 mm (2 in.) from the popped nail.
2. Hammer the popped nail into the wall. Fill dents with drywall compound. If drywall screws were used, countersink the screw (so that it rests slightly below the surface) using a star type (Phillips) screwdriver. In order for the nail or screw to hold, the head should not break the surface of the wallboard paper.
3. When the compound is dry, sand lightly, paint with a primer coat, then paint to match the wall.
4. Other small holes can be filled with compound, sanded lightly, then painted.

Note: To reduce the amount of sanding needed, lightly wipe over the compound you applied with a damp sponge.

Repair a large, damaged area or hole

1. Use a straightedge or square to

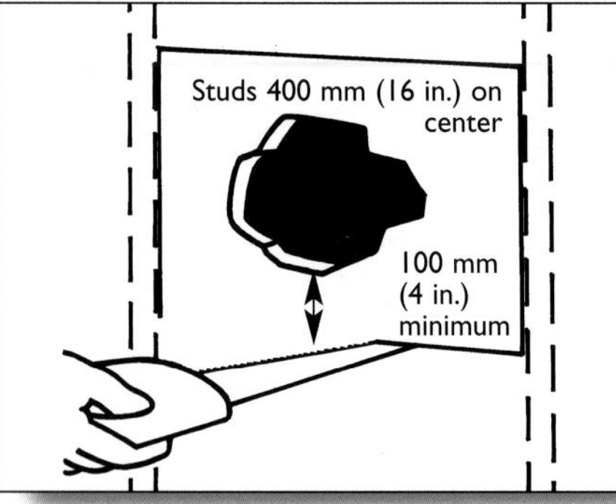

outline the damaged area.

2. Use a utility knife or drywall saw to cut around the outlined area and remove the damaged section. Avoid damaging the air and vapour barrier if the wall is an exterior wall. Cut one or more pieces of 12 mm (1/2 in.) plywood to slide into the hole, partly behind the edges of the surrounding drywall. Screw through the surrounding drywall to hold the plywood in place. The plywood creates a backing, where required, for the patch.

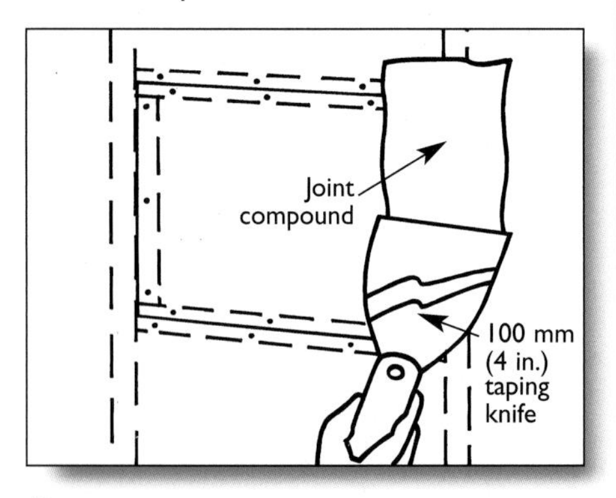

3. Cut a wallboard patch to fit the hole. Screw the patch to the installed backing or studs.

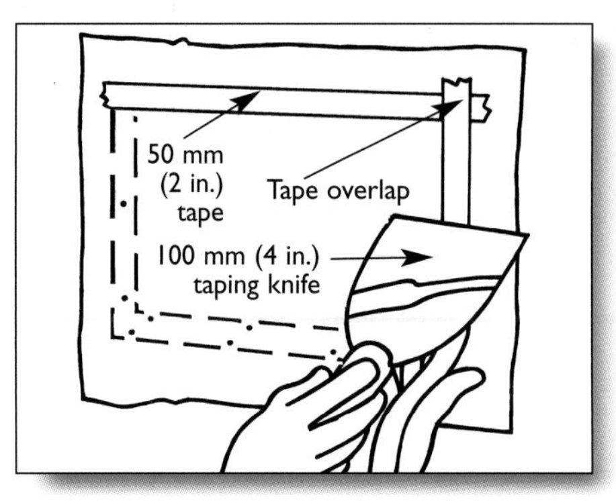

4. Cut pieces of glass fibre drywall tape and apply over the edges of the patch. Alternatively, apply a thin layer of joint compound over the edges and embed paper tape into the compound.

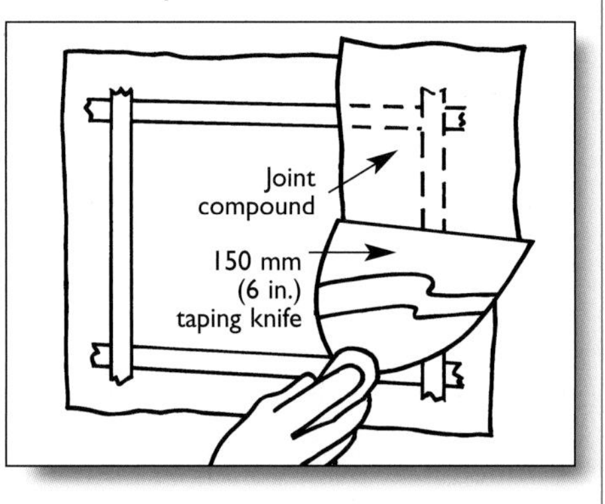

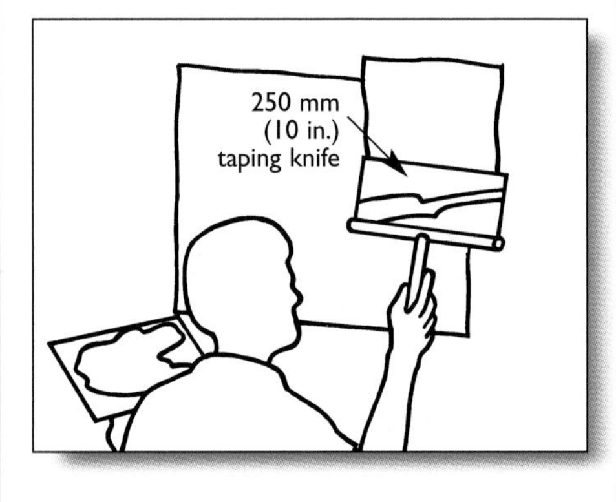

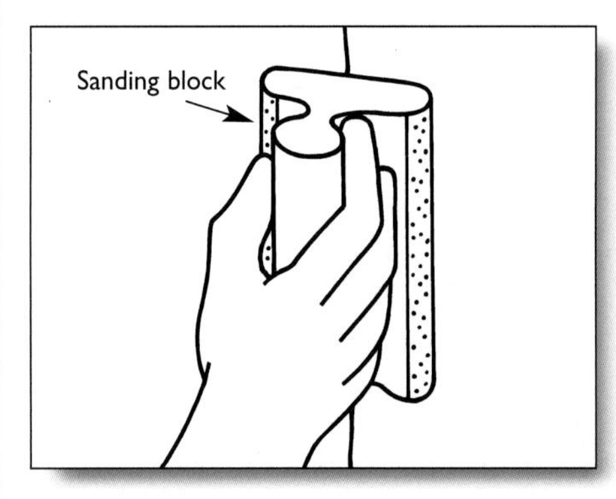

5. Cover the glass fibre or paper tape with a layer of wallboard compound, using a drywall trowel. When the compound is dry, apply a second coat over a larger area and feather the edges (compound layers gradually get thinner at the edges). After the compound dries, sand the area lightly until smooth. Repaint the area to match the wall.

Fill and finish surface cracks

Narrow cracks can occur in walls and ceilings due to normal movement of the building from settling and changes in temperature and humidity.

Skill level rating: 2 - Handy homeowner

Materials: premixed wallboard compound, paint

Tools: wallboard knives or scrapers, medium grit (80-100) sandpaper, small block of wood or sanding block, old cloth or paintbrush

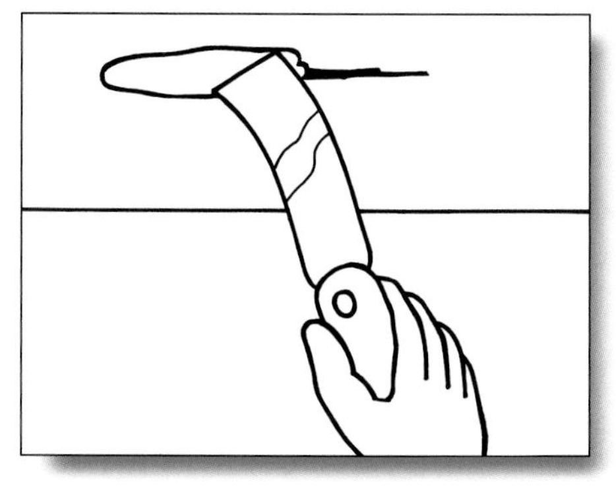

1. Fill cracks with wallboard compound.

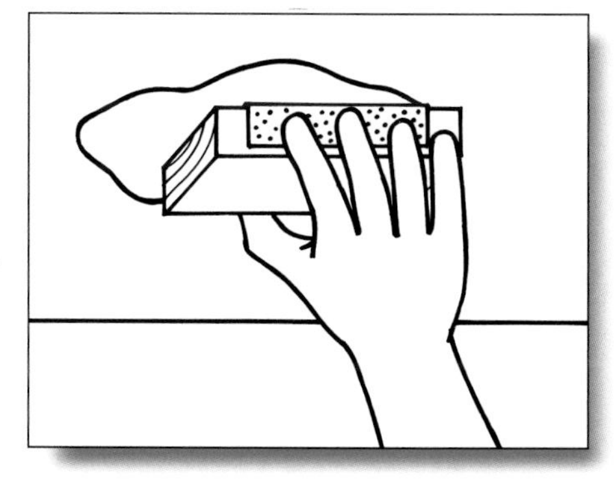

2. When the compound is dry, sand lightly, paint with a primer coat, then paint to match the wall.
3. For larger cracks, remove any loose material and smooth the edges of the crack with a utility knife. Apply glass fibre mesh tape or paper tape and finish as described above.

Repair drywall taping that has pulled away from the wall

Drywall tape sometimes pulls away from the wall and needs to be repaired. This is usually caused by improper initial installation.

Skill level rating: 2 - Handy homeowner

Materials: premixed wallboard compound, paper or glass fibre mesh tape, paint

Tools: utility knife, wallboard knives or scrapers, medium grit (80-100) sandpaper, small block of wood or sanding block, old cloth or paintbrush

1. Strip out any detached tape. With a utility knife, cut the tape where the detached portion ends.
2. Clean any loose joint compound from the affected area.
3. Apply a thin coat of premixed wallboard compound over the affected area. Embed a new strip of paper tape. Alternatively, apply glass fibre mesh tape directly over the affected area before applying any joint compound.
4. Cover the glass fibre or paper tape with a layer of wallboard compound, using a wallboard trowel. When the compound is dry, apply a second coat over a larger area and feather the edges (compound layers gradually getting thinner at the edges). After the compound dries, sand the area lightly until smooth. Repaint the area to match the wall.

Caulk and fill cracks around the bathtub or shower

Loose grout or sealant usually causes cracks around the bathtub, shower or bathroom tile joints. If these cracks are not quickly cleaned out and filled, they can let water in that will damage your walls and the framing behind them, providing a perfect environment for mold growth inside your walls. Cracks also catch dust and grow mold. Dust and mold are increasingly being recognized as sources for health and indoor air quality problems.

New grout should be installed in joints between tiles. Since tubs expand and contract slightly due to temperature differences, grout is not suitable between the tub and the tile walls. Silicone sealant is required in that location.

Skill level rating: 2 - Handy homeowner

Materials: masking tape, waterproof grout or bath and tub silicone sealant

Note: Grout comes in pre-mixed or powder form that you mix with water. Grout is harder to work with than silicone sealant, but is less expensive. Silicone sealant comes in either squeezable plastic tubes or in larger cylindrical tubes designed for use in caulking guns. Be sure to follow the directions on the label before using either product.

Tools: cold chisel or slotted screwdriver, putty knife, bowl or caulking gun, rubber gloves

1. Scrape the old grout or sealant from the crack, using the edge of a chisel or an old slotted screwdriver. Protect the finish on the bathtub by putting masking tape on the edge of the tub, close to where you will be working and a drop cloth to cover the remainder.

2. Thoroughly clean the crack to remove any soap, grease, or dirt. To use grout, keep the crack damp and follow the mixing directions on the package. Work the mixture into the crack to fill it completely, using the putty knife. Smooth the surface. Wipe away any excess grout with a damp cloth before it gets hard. Allow the remaining grout to dry completely (usually 24 hours) before using the tub or shower. Do a finishing polish to remove any grout residue.

Note: Grout is very tough to remove once it has hardened. Don't wash leftover grout down the sink, as it may harden there and block the drain! Dispose of it in your regular garbage instead, and wash all traces of grout off the bowl and putty knife before putting them away.

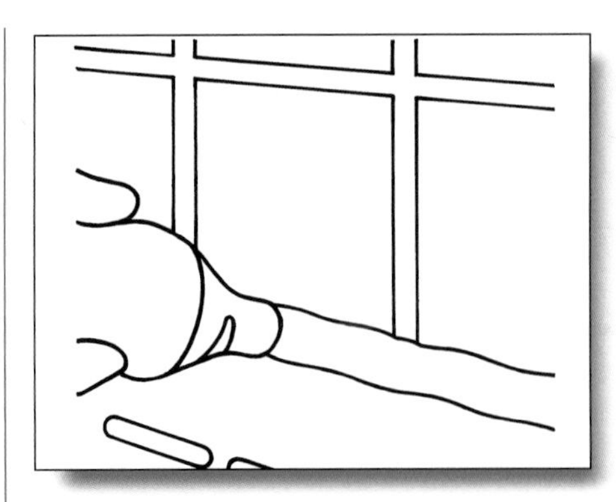

3. To use silicone, let the crack dry completely first. Then cut the end of the nozzle on the sealant tube at an angle so the opening will match the width of the crack. Apply a strip of masking tape along each edge of the crack to create the desired width of caulking bead. Puncture the seal at the bottom of the nozzle and mount the tube in a caulking gun if required. Hold the nozzle at a 45-degree angle in contact with both edges of the work. Force a steady bead of sealant into the crack. Try to gauge the amount of sealant required to create a smooth bead. If necessary, remove any excess sealant with a dry paper towel. Smooth the surface by wetting your gloved index finger and running it along the filled crack. You have to act fast, as silicone "skins" or surface dries in only five to seven minutes. Complete drying usually takes at least a day.

Note: Silicone sealant, before it fully dries, can be irritating to skin and eyes, and must be used with the closest window open for ventilation.

Replace ceramic tiles

If you notice a loose or missing ceramic wall tile, it should be repaired or replaced as soon as possible to prevent further damage. Choose a solvent-free adhesive. The label on the product should say solvent-free or non-toxic.

Skill level rating: 2 - Handy homeowner

Materials: grout , tile adhesive, tile

Tools: pencil or non-permanent marker, chisel, mixing bowl, utility knife, nailset, combination square or straight edge, pliers, measuring tape, scrap piece of wood to trim tiles on, tile adhesive applicator, rubber gloves, eye protection, nippers, tile cutters.

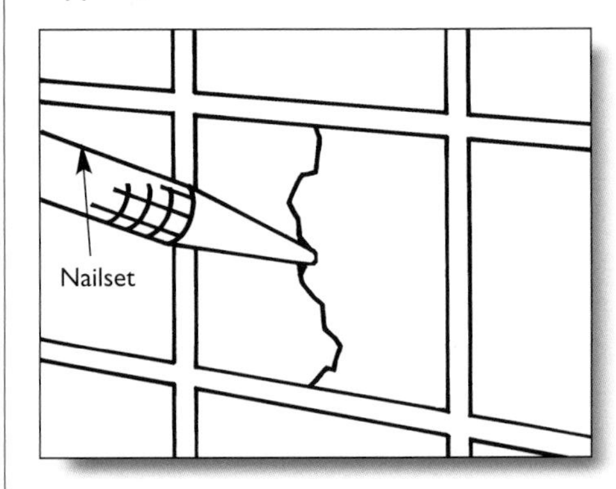

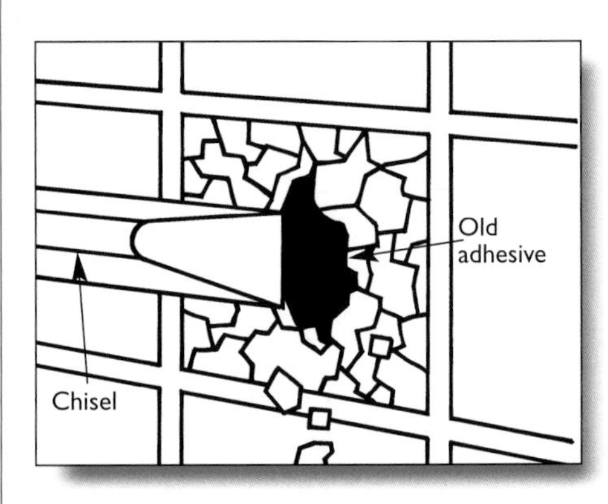

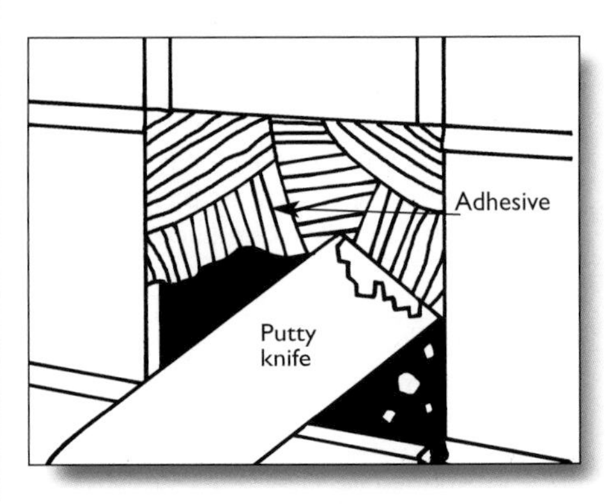

1. To remove an old tile first scrap out grout with the edge of a cold chisel or old slot screwdriver, then remove tile or if necessary, break the tile into small pieces with a hammer. Wear protective eyewear to protect your eyes from flying tile. Scrape the old adhesive and other loose material off the wall. If the wall where the new tile will be installed is damaged from pulling off the old tile, level and seal it with a thin layer of adhesive. Allow to dry. All used tiles that will be reinstalled should be as clean as possible on the back side.
2. If you are using a new tile, you may have to cut it to fit. Mark the cut line or lines on the top (glossy) surface of the tile using a pencil or non-permanent marker. Make straight cuts by lining up a straightedge or square with the marked lines and scoring (a cut on the tile that just cuts through the top surface) the surface with a glass cutter or sharp utility knife. You must press down heavily enough to cut through the tile glaze, so keep your fingers clear of the blade. Afterwards, place the scored line over the edge of a table top, press the tile down firmly with the palm of one hand, and snap the overlapping edge off with the other hand. The tile should break cleanly off at the line. If you've never done this before, try a practice piece first. Or, if possible, rent and use a tile cutter.

3. Cut curved or round lines as a series of short interlocking straight scores made just inside the area perimeter. Score the enclosed area with hatches, then carefully nip out the enclosed, hatched area with pliers. Rough edges can be ground to shape with an inexpensive grinding bit mounted on a power drill. Nippers are available to make this task much easier.
4. Test fit the tile, then spread ceramic tile adhesive on the space to be filled. Press the tile or tiles firmly in place, in line with the existing tile edges.

5. Once the tile has set firmly, (refer to the manufacturer's instructions for the tile adhesive) fill the joints with grout. Mix only as much as you need, adding water to the powder to form a paste. Press the mixture into the joints with a putty knife and finish tooling the joints with a wet, gloved finger.
6. Let grout dry for about an hour. Wipe excess grout from tiles and other surfaces. Wash grout off tools in a shallow pail of water.
7. Dispose of leftover grout in your garbage, not down your sinks!
8. Keep the newly grouted joints dry overnight. Do a final polish to remove any grout residue.

Reattach items such as towel bars, handrails, curtain rods that have pulled away from the wall

Sometimes towel bars, handrails or curtain rods pull away from the wall, especially if they were not well fastened originally. If the screws through the attachment brackets penetrated into a stud, it may be possible to use a larger screw in the same location. However, when using a longer screw, there is always the danger of puncturing hidden wires or pipes. It is often better to move the bracket a little to fasten into solid wood.

The same is true when fastening through drywall alone. The original attachment location is probably damaged and unsuitable for reattaching the bracket.

Skill level rating: 2 - Handy homeowner

Materials: toggle bolt or other hollow wall fastener

Tools: drill, screwdriver, measuring tape, stud finder

1. Remove the loose bracket, if necessary.
2. Examine the original fastener anchor or use a nail to probe into the original anchor hole to try to determine if fastening was done to a stud or hollow wall.
3. A longer fastener may be enough to reattach the object securely. If not, determine another suitable location for the bracket. Hold the bracket in the planned location and mark the screw holes. Use a finish nail to lightly probe the new location to determine whether you will be fastening into a stud or hollow wall (stud is preferred).
4. Patch the damaged location following the appropriate wall patching procedure *Patching holes*.
5. Drill the correct size holes in the new locations for the screws or hollow wall anchors.
6. Insert the wall anchors if required.
7. Screw through the brackets into the stud or wall anchors.
8. Reattach the handrail or curtain rod.

Seal gaps around plumbing vents that penetrate into attic or through exterior walls

To operate properly, all plumbing drain lines must be vented. The vent stacks run typically from the plumbing drain lines up through the ceiling and house roof, allowing outside air to replace water being flushed or drained away. Sometimes, plumbing traps from kitchen or bath sinks are vented vertically through an exterior wall cavity to connect to a main vent. Without vents, the draining action would create a vacuum and pull water and air through plumbing traps. Without the traps, sewer gases could back up into the house.

Usually, there is at least one large diameter, 76 mm (3 in.) vent stack for each toilet and smaller vents that run from other plumbing fixtures and connect to one of the larger stacks. In many houses, the holes through the ceiling or walls for the plumbing vents are not sealed properly. Any attic penetrations may allow leakage of warm, moist house air into the attic, which is a major cause of attic moisture. Improperly sealed penetrations through the interior air/vapour barrier of exterior walls may allow air and moisture leakage that can be uncomfortable and cause damage to the walls. All ceiling and wall penetrations should be tightly sealed.

Because the ABS plastic vent stacks (or even the older cast-iron stacks) are subject to warm moist air as well as cold air from outside, they may expand or contract slightly along their length. A flexible seal works best. Since most vent stacks are routed inside frame walls, access to the stack is usually easiest from the attic.

Skill level rating: 2 - Handy homeowner

Materials: rubber gasket material or vent stack roof flashing ("roof boot"), non-hardening acoustical sealant, low expansion closed cell polyurethane foam, contractor's sheathing tape, roofing nails

Tools: utility knife, caulking gun, hammer, dust mask, goggles

Attic penetration

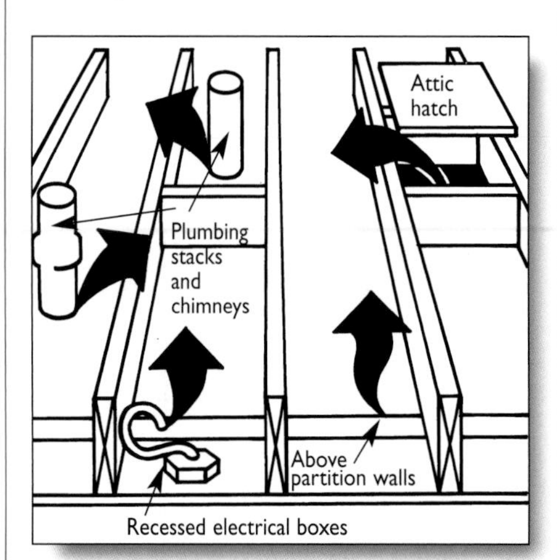

1. Enter the attic and locate the plumbing vent stack where it comes up through the ceiling. Be careful where you step when you are in the attic. Walk only on trusses or joists so that you do not fall through the ceiling below.
2. Move insulation away from the stack to allow room to work. Be careful to wear respiratory and eye protection when disturbing insulation.
3. If installing a flexible seal is possible, prepare a piece of sheet rubber gasket about 300 mm (12 in.) square with a round hole in the centre to accommodate the pipe and a slit from the edge to the hole (or, for the right size pipe, use a rubber roof boot with a slit).
4. Fit the rubber gasket around the pipe.
5. Run a bead of acoustical sealant under the edge of the gasket against the top of the polyethylene air/vapour barrier or the ceiling material if there is no polyethylene sheeting.
6. Use roofing nails to fasten the rubber sheet to the top of the wall plate, if possible.
7. Tape the edge of the rubber gasket to the top of the polyethylene air/vapour barrier or ceiling material.
8. Run a bead of acoustical sealant along the slit in the gasket from the edge to the centre.
9. In the rare situations where expansion is not expected, it is possible to seal the ceiling penetration around the vent stack with low expansion polyurethane foam instead of using a flexible gasket. Follow the manufacturer's installation instructions.
10. Replace the ceiling insulation.

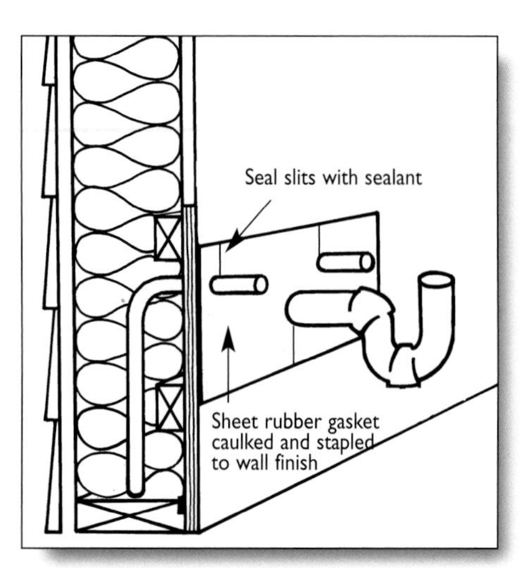

Exterior wall penetration

1. If installing a flexible seal is possible, prepare a piece of sheet rubber gasket about 150-200 mm (6-8 in.) square with a round hole in the centre to accommodate the pipe, and a slit from the edge to the hole.
2. Fit the rubber gasket around the pipe.
3. Run a bead of acoustical sealant under the edge of the gasket against the polyethylene air/vapour barrier or the wall finish.
4. Tape the edge of the rubber gasket to the polyethylene air/vapour barrier or wall finish.
5. Use staples and thin wood strips to fasten the rubber sheet to the wall, if possible.
6. Run a bead of acoustical sealant along the slit from the edge to the centre.
7. Alternatively, instead of using a rubber gasket, seal the penetration around the vent with low expansion polyurethane foam or caulking. Follow the manufacturer's installation instructions.

Seal wall and ceiling electrical boxes and wires

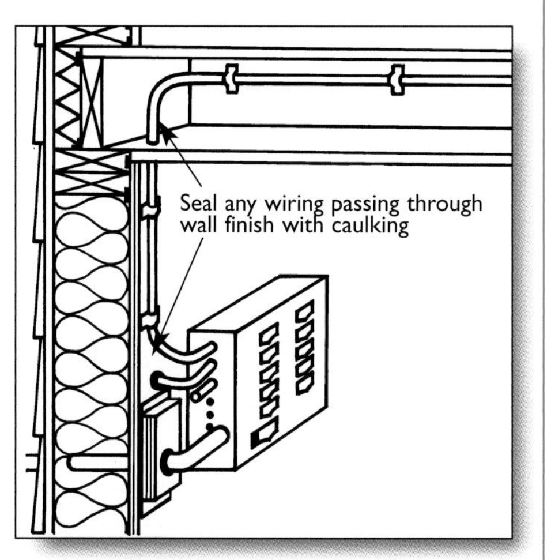

In many houses, there is only insulation, not a sealed air/vapour barrier above metal ceiling electrical boxes or behind wall boxes. Holes for the passage of wires through wall top plates are often not sealed at all. Insulation does little to prevent air leakage. Since there may be a number of these penetrations, a lot of warm, moist house air may leak into the exterior walls or attic causing not only heat loss but also moisture damage.

Air leakage occurs through holes, not directly through finished wall or ceiling materials like drywall or plaster. All wall or ceiling penetrations should be tightly sealed.

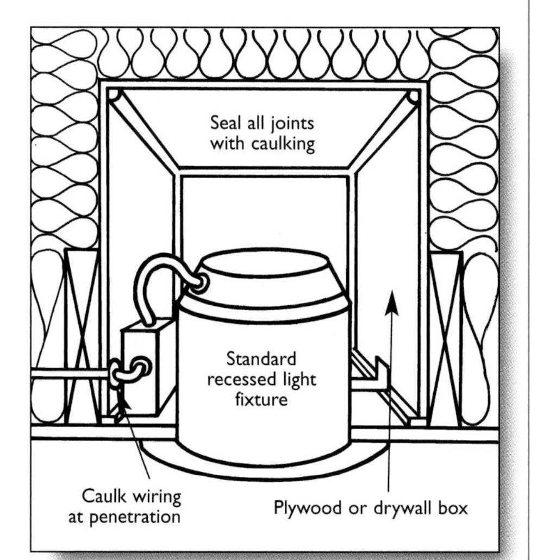

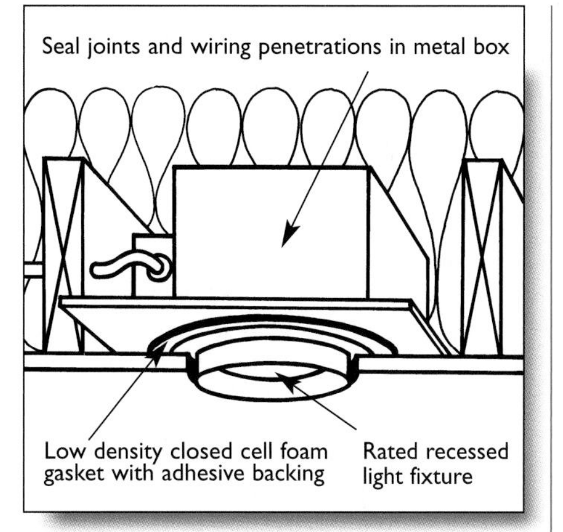

Caution: Insulation should not be placed around recessed ceiling light fixtures, unless the fixture is specifically designed to accommodate the insulation. Overheating and a fire hazard may result. If recessed lighting fixtures are CSA rated to be covered by insulation and are accepted by the local electrical inspector, then any holes in the metal box may be sealed with aluminum foil duct tape and insulation placed over the box. If the recessed fixture is not appropriate for direct insulation cover, follow the manufacturer's instructions and building code requirements regarding clearance around it. A manufactured plastic box or a site-built plywood box that provides the correct clearance may be used to air seal around the fixture.

Skill level rating: 2 - Handy homeowner

Materials: polyethylene air/vapour barrier, acoustical sealant, contractor's sheathing tape, receptacle box gaskets

Tools: utility knife, caulking gun

1. Enter the attic and move sections of insulation one at a time to locate any ceiling penetrations. There will be plumbing vent stacks (described above), ceiling electrical boxes and wires that penetrate the top plates of walls. Wear respiratory and eye protection when disturbing insulation.

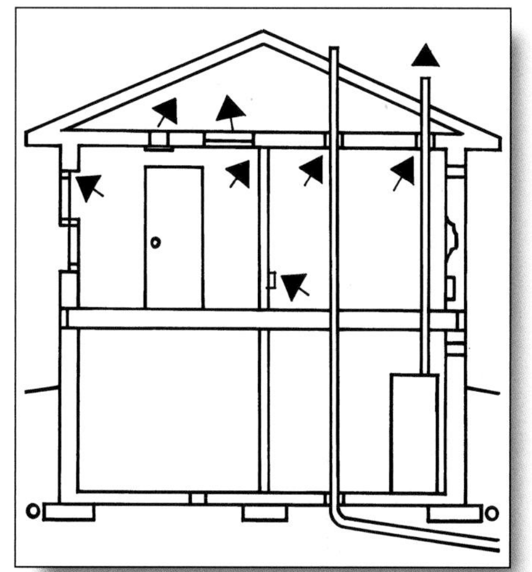

2. Using polyethylene sheeting, place a patch about 300 mm square (12 in.), over any unsealed electrical box.
3. Tape the patch to the top of the ceiling air/vapour barrier or ceiling finish. (Before taping, make sure that the surface is clean so that the tape will stick well.)
4. Use tape or acoustical sealant as required to seal the patch to the ceiling and to seal any wires to the patch.
5. Use tape or acoustical sealant to seal any holes around wires that penetrate the top plates of walls.
6. Finally, install manufactured foam gaskets behind the cover plates of all exterior wall receptacles.

Painting the interior

Painting your house regularly not only makes it look fresh and clean, it prevents damage that could later cost money. Peeling paint or surfaces that have never been protected with paint can lead to rotted and unsightly woodwork. Prevention is cheaper than repairing the damage after it is done.

Skill level rating: 2 - Handy homeowner

Materials: paint

Tools: rollers and tray, extension handle for the roller (if you are painting ceilings or high walls), paintbrushes 50 mm (2 in.) to 100 mm (4 in.) wide, preferably with chisel edges, an angular sash brush (good if you are painting window frames, mouldings or other narrow surfaces), detergent, sponges or cleaning cloths, masking tape, drop cloths, stepladder, paint bucket, mixing paddles, protective clothing

Prepare to Paint

- Consider whether the old paint may have lead in it. Refer to the *Safety tips* earlier in this section.
- Before you paint or repaint, check the surface carefully and repair any defects such as crayon marks and nail holes.
- Use drop cloths or plastic sheets to protect your floors and furniture against paint splatters.
- Remove curtains, pictures, electrical switch plates and outlet plates. Cover the switches and receptacles with masking tape. If other fixtures cannot be removed, edge them with masking tape. Be careful not to get wet paint into electrical openings.
- Remove small pieces of furniture. Larger pieces may be pushed into the middle of the room and covered with newspapers or a plastic drop sheet.
- Exposed nail heads should be hammered in and countersunk, then filled with patching compound, sanded smooth, and wiped clean before painting.
- Fix popped nails and cracks in wallboard. Fix cracks and holes in plaster. Refer to *Patch small holes* earlier in this section.
- Clean dirt, smoke and grime off walls and ceilings.
- Scrub moldy areas with a mild unscented detergent solution. Rinse with clean water and let dry. Refer to *Clean up visible mold* in the *Foundations and Basements* section.
- Sand glossy areas lightly so that the fresh paint can adhere.
- Wash away loose, powdery, flaking or peeling paint with unscented detergent and warm water. Loose, flaking or peeling paint must be scraped off and the depression filled and sanded smooth. Spot prime with the paint you have chosen.
- Remove old wallpaper and paste by soaking with warm water or special mixtures made for the job before scraping them off. Wash away excess glue before painting.
- Unpainted plaster, drywall, concrete and cement blocks should all be thoroughly dry and aged or cured before painting. They should be primed with a sealer coat first.
- Seal knots in unpainted wood with an enamel undercoat.

How Much Paint to Buy

The quantity of paint required depends on the surface being painted and the type of paint used. A safe estimate would be to allow 1 L (about 1 quart) of paint to cover approximately 8 m^2 (about 86 ft^2) of wall surface for each coat. Several coats may be required depending on the quality of the paint and the colour being covered.

Apply the Paint

Assemble all the tools that you need. Work in comfortable clothes at a comfortable room temperature—20°C to 22°C (68°F to 72°F) is ideal—and work at a comfortable speed. Provide ventilation, preferably by opening the windows before you paint. Always allow the full drying time between coats and remember that temperature and humidity can affect drying times. Follow the instructions on the paint can.

Painting Sequence

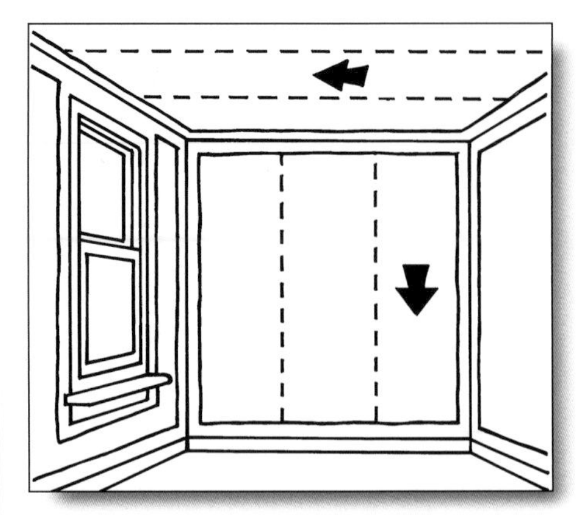

Ceiling: Start with the ceiling if you are painting both walls and ceiling. Begin in a corner and work in small areas as far as your arm can comfortably reach. Paint across the shorter measure of the ceiling.

Walls: Work in small convenient areas, painting from top to bottom. With a 50 mm (2 in.) trim brush, "cut in" the ceiling and wall edges, then use a roller to fill in.

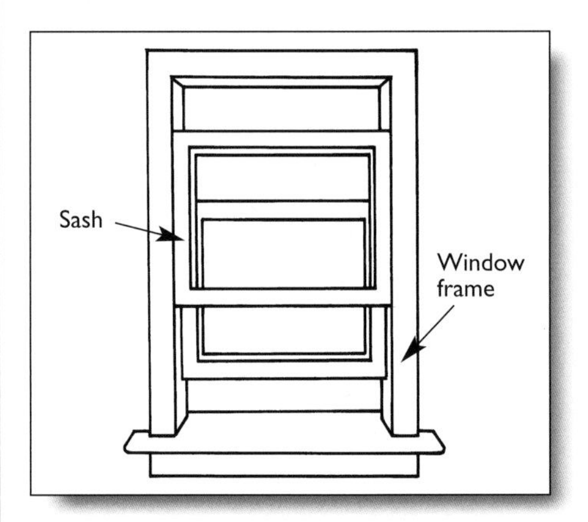

Windows: These need only patience and a steady hand. A 50 mm (2 in.) angular sash brush lets you apply the proper amount of paint. Paint the window sash top, front and muntin bars (the narrow bars between the frames) first. Finish by painting the window frame, top, sides and bottom.

Doors: Paint door panels first, then horizontal strips, followed by vertical strips. If the door opens into the room being painted, paint the latch edge. If the door does not open into the room, paint the hinge edge. Note: You can help to prevent sticking and dragging by sanding door edges before painting. This will keep paint from building up and causing the door to stick.

Baseboards: Coat the top of the baseboard first, then the floor edge. Use masking tape to protect varnished or carpeted floors.

Cupboards: Remove all drawers and hardware. Paint the inside of the cupboards first. Paint moldings around any trim panels. Paint the rest of the frame, then the top. On the drawers, paint only the exposed edges and the front panels. Stack drawers, bottom down, until thoroughly dry.

PLUMBING

Plumbing systems usually work well if they are maintained regularly, small problems are fixed promptly and simple measures are taken to prevent problems. Most plumbing problems that do occur involve leaks or clogs in fixtures or pipes.

The most common plumbing problems are:

- leaking faucets
- clogged drains
- overflowing toilets
- toilets that run constantly
- toilet flushing problems
- condensation on tanks
- leaking toilets at the base
- leaking shower stalls and tub surrounds
- plumbing leaks in all types of pipes
- plumbing noises
- frozen pipes
- problems with water pumps and pressure tanks

Maintenance includes:

- Hot water heaters—flushing the tank and testing relief valves
- Septic system—periodic pumping, practices to avoid overload or damage to the system
- Water treatment equipment—changing filters and adding chemicals
- Sump pumps—checking to ensure the pump and discharge lines operate.

Prevention tips

- Prevent clogged drains. Dispose of grease, hair or food in the garbage, not down the drain. Use a strainer basket in your sink to catch food bits and pieces.
- Use a homemade drain cleaner once a month to clean out kitchen and bathroom drains. Do not use chemical drain cleaners.

Homemade drain cleaner

Once a month or so, it's a good idea to clean out the kitchen and bathroom drains. Use a homemade cleaner, which is economical and will not damage your plumbing and septic system.
Ingredients (Buy these at the grocery store)
250 ml (1 cup) baking soda
250 ml (1 cup) table salt
65 ml (1/4 cup) cream of tartar
Mix dry ingredients in a bowl. Store it safely out of children's reach, in an empty jar or tin and label clearly.

To use
Put about a 65 ml (1/4 cup) of the mixture into the drain and add one cup of water. The mixture will fizz and bubble. When the bubbling stops, run clear water through the drain.

- Prevent clogged toilets. Throw disposable diapers, sanitary napkins, wads of tissue and cigarette butts in the garbage, not in the toilet.
- Improve water flow by keeping the screens cleaned in faucets.
- Do not store things on or against the hot water tank, particularly near the exhaust hood or pilot light in the case of a gas or oil heater. Do not store in this area any items that may give off combustible fumes and could cause an explosion, including gasoline, paints and cleaning products.
- Do not let anything fall against the hot water tank. Any blow or sudden jolt could crack the glass lining inside.
- Insulate your hot water tank with a hot water tank blanket, available at your hardware or plumbing store. These special insulating covers help hold the heat in the tank and conserve energy. When insulating any fuel-fired hot water tanks, it is very important not to insulate over any controls or to obstruct combustion air openings or vent connections. The insulation should not come in contact with the vent connection. Before insulating, check with your local installer or gas utility to ensure that you will not compromise the safety or operation of the water heater.
- Ensure that all occupants know where the main water supply shut-off valve is located. Shut-off valves for the main water supply and fixtures should be closed and opened periodically to ensure that they are not stuck in the open position. Both the main valve and fixtures valves must be operable so that water can be turned off in an emergency or when plumbing repairs are necessary.

Repair tips

- Before starting any plumbing repairs, shut off the water as close to the repair area as possible. Depending on the repair, you may also need to first drain the pipes.
- Protect chrome taps with masking tape, painter's tape or a piece of rubber tubing (that is, bicycle inner tube) to avoid damage.
- Even though a name and model number may be all you need so that you ask for the right replacement part, it's always best to take the parts that have to be replaced with you to the store. This way, you'll get exactly what you need.
- Always follow the manufacturer's instructions for the specific fixture that you're installing.
- Rent specialty tools such as a toilet auger if you need it.

Special considerations

Healthy Housing™

- Leaking faucets, toilets, and pipes waste water. One drip per second from a leaking hot water tap or shower head sends about 800 L (about 200 gal.) of hot water down the drain every month. That's not only water going down the drain—it's money, whether it's for your water bill, hot water bill or more frequent pumping needed for a septic system. The leaking water also keeps surrounding areas moist, which can cause iron stains and lead to mold growth.

- Installing aerators in your faucets is an easy and very inexpensive maintenance task that can reduce the amount of water you use by half or more. If a faucet, shower head or toilet needs to be replaced, choose low flow fixtures that will reduce the amount of water you use in your home. During a six-minute shower, using a low-flow shower head, you could save as much as 108 L (23 gal.).

- If your toilet is old and needs repairs, it may be better to replace it with a low flush toilet with an insulated tank. These toilets use only 6 L (1.3 gal.) of water per flush instead of older toilets that use 20 L (4.3 gal.) or more. The insulated tank will prevent "tank sweating" that can cause mold growth on the drywall behind the tank or on surfaces where the moisture drips.
- Chemical drain cleaners are dangerous to use and pose serious hazards to the environment.

Safety

- Protect yourself from serious health risks due to lead exposure. Use lead-free solder when soldering copper pipes.
- Regular water testing for bacteria is recommended for households that depend upon well water. For a drilled well, test the water twice a year—in the spring and fall. If the water is from a dug well, test the water three times a year—in the spring after the snow melts, mid-summer and fall. Contact your local health unit or health department to find out where test bottles are available. In many areas, testing is free and the sample bottles are available at the health unit.

Tasks

Repair leaky faucets

There are two general types of faucets—faucets that have washers (compression faucets) and washerless faucets (ball, ceramic disc or cartridge). Faucets may look different, but they usually leak because they have worn or cracked washers, seals, O-rings or cartridges. Replacement parts are available at hardware or building supply stores.

Compression faucet

Skill level rating: 3 - Skilled homeowner

Materials: assorted plumbing washers, packing wicking, waterproof grease

Tools: adjustable wrench or pliers, screwdriver, utility knife.

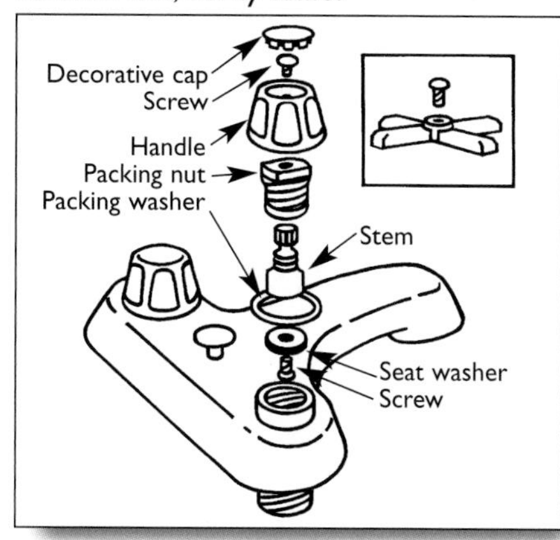

1. Note whether the water is leaking from the tap spout or under the handle of the faucet. Turn off the water at the nearest shut-off valve. Next, turn on the tap and wait until the water stops running.
2. Unscrew and remove the handle. The screw to remove the handle may be underneath the cap on the top of the faucet.
3. If you suspect that water was leaking around the packing or cartridge nut under the handle, try tightening the nut with a wrench. Test the tap and if water still leaks, turn off the water again and loosen the nut.
4. If this is not the cause of the leak loosen the packing or cartridge nut and pull out the tap's valve unit or cartridge.
5. Remove the screw holding the old washer at the end of the stem and replace with a washer of the same size and type. Replace the washer screw and any O-rings if they show signs of wear. Coat new washers and o-rings with grease.

Note: Very old fixtures may have wicking or a packing washer instead of a rubber washer. If your tap has packing, wrap the spindle with packing wicking under packing nut.

6. Replace the valve unit and turn it to the open position with the unattached handle. Tighten the packing nut, then close the valve and reattach the handle.
7. Turn the water back on and test for leaks.

Ball faucet

Materials: repair kit for your particular faucet

Tools: adjustable wrench or pliers, screwdriver, utility knife

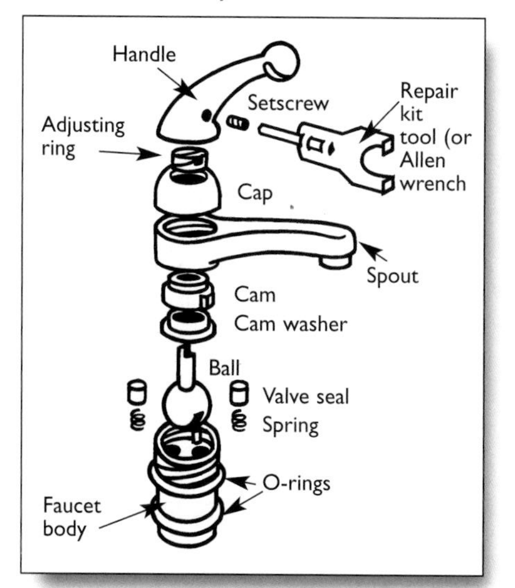

1. Turn off the water at the nearest shut-off valves. Turn on the tap and wait until the water stops running. Loosen the setscrew holding the handle. Remove the handle and inspect the adjusting ring for dirt and wear.
2. Tighten the adjusting ring and test to see if this has fixed the leak. If not, turn off the water again, run the water until it stops and unscrew the cap to the faucet body. Remove the cam, cam washer, and rotating ball.
3. Gently remove the valve seals and springs inside the faucet.
4. Remove the faucet spout by gently pulling upward. Cut off O-rings and replace with new ones from the repair kit. Coat new washers and O-rings with grease. Reinstall the spout.
5. Replace springs, valve seats, ball, cam washer and cam with new parts from repair kit.
6. Reinstall faucet cap and faucet handle.
7. Turn the water back on and test for leaks.

Cartridge faucet

Materials: replacement cartridge for your particular faucet, waterproof grease.

Tools: adjustable wrench or pliers, needle-nose pliers, screwdriver, utility knife.

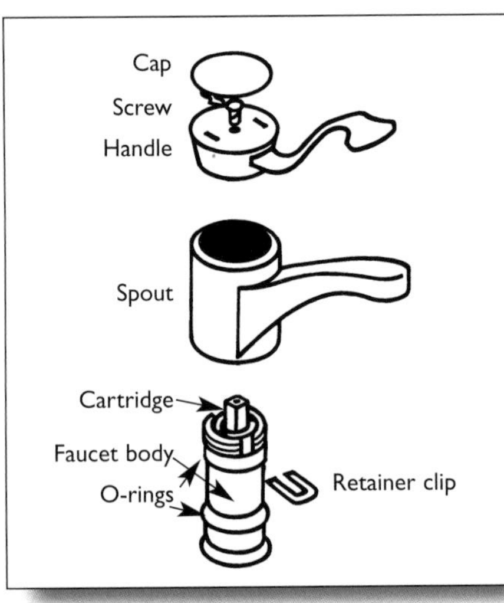

1. Turn off the water at the nearest shut-off valves. Turn on the faucet and wait until the water stops running. Remove cover cap and unscrew handle screw. Raise lever to allow it to release from the stem and lift off.
2. Remove the retaining nut from the faucet stem.
3. Pull out the retaining clip from the top of the cartridge.
4. Pull the cartridge straight up using pliers.
5. Replace the cartridge with a new cartridge of the same size and type. Re-insert the retaining clip.
6. While the faucet is apart, you may want to replace the O-rings for the spout. Pull the spout up and off to expose the tap stem. Remove the old O-rings and replace with new rings of the same size and type. Coat new O-rings with grease. Replace the spout.
7. Replace the retaining nut, collar, handle, handle screw and cap.
8. Turn the water back on and test for leaks.

Ceramic disc faucet

Materials: replacement cartridge for your particular tap

Tools: adjustable wrench or pliers, needle-nose pliers, screwdriver

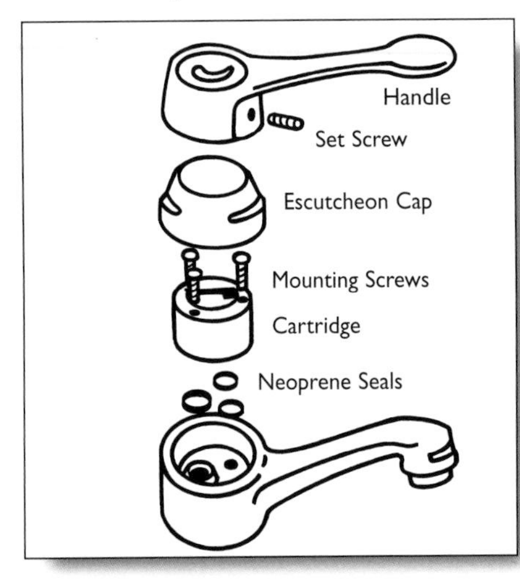

1. Turn off the water at the nearest shut-off valves. Turn on the faucet and wait until the water stops running. Remove cover cap and unscrew handle screw. Remove handle.
2. Continue to disassemble parts and carefully place them in sequence to make it easy to re-assemble. Remove the cap, and cartridge assembly.
3. Replace any worn inlet seals and the cartridge.
4. Reassemble the parts in sequence or referring to manufacturer's instructions.
5. Turn the water back on and test for leaks.

Clear clogged sink or tub drain

When a drain clogs, you must find out whether the clog is in the one drain only or in the main drain. By running water through the drain in another part of your home, you can soon find out. Don't flush a toilet to try to locate the stoppage. Flushing the toilet releases too much water and may cause an overflow. Instead, half fill a sink in another part of the house. Pull the plug. If the water drains easily it means the clog is only in the drain under the stopped-up sink or tub. If the water does not run out, it means the main drain that connects all your sinks and tub is plugged. You can't fix this problem yourself. Call the plumber.

You can clear a single clogged drain by:

- cleaning the strainer
- cleaning the stopper
- using a plunger
- cleaning out the trap
- using a drain auger.

Skill level rating: 2 - Handy homeowner

Skill level rating: 4 - Qualified tradesperson/contractor (plumber) will be needed if the mechanical methods don't work.

Materials: water

Tools: screwdriver, plunger, wrench, groove-joint pliers, needle-nose pliers, funnel, drain auger, coat hanger, small container to catch water

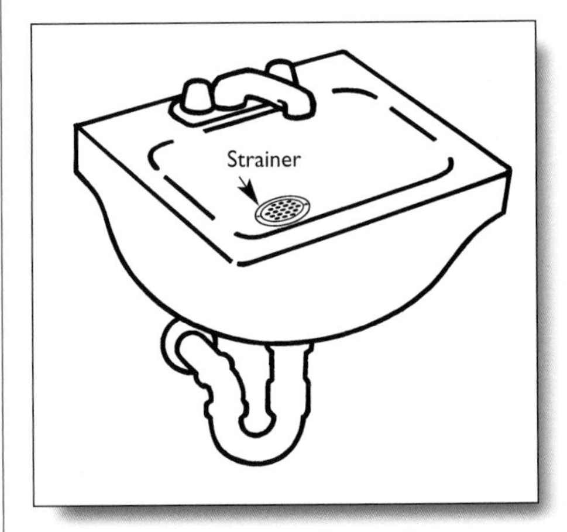

Clean the sink or tub strainer

1. If your sink or tub has a removable strainer, remove the screw from the strainer to a safe place.
2. Lift the strainer up and out of the sink.
3. Clean the strainer and the sink opening below (tailpiece) as far as you can reach. Needle-nose pliers are handy for removing hair in the drain. A coat hanger can also be used to reach down further.
4. Run water through the sink to see if it drains properly. If so, replace the strainer and screw it down.

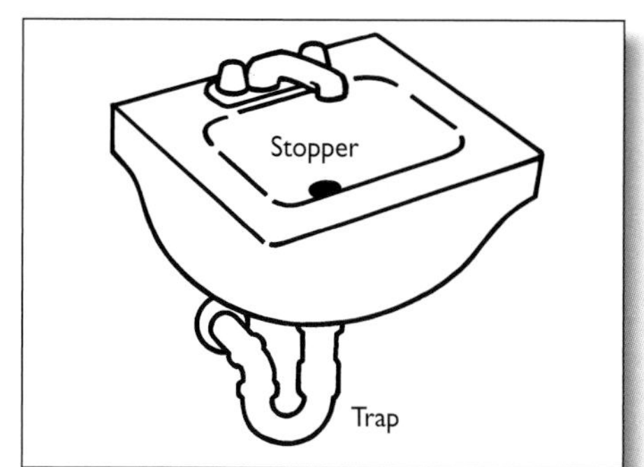

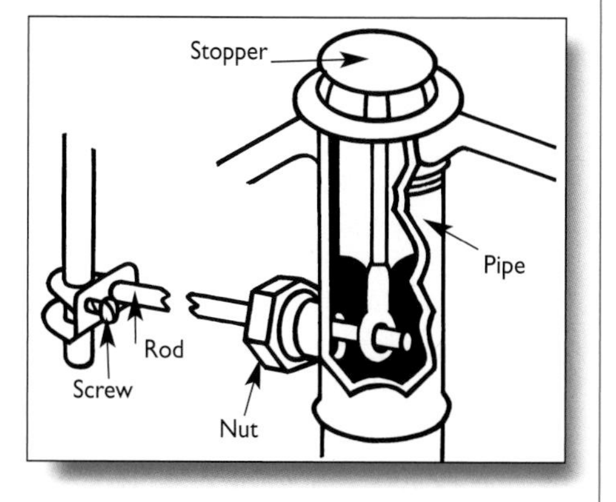

Clean the sink or tub stopper

1. If your sink or tub has a stopper, try to remove it by turning it counter-clockwise and lifting.
2. If the stopper cannot be removed by turning, unscrew and remove the rod holding it back. To do this, place a container under the sink, loosen the screw and remove the nut from the pipe. You should be able to lift the stopper out.
3. Clean the stopper and the drain opening as far as you can reach. Needle-nose pliers are handy for removing hair in the drain. A coat hanger can also be used to reach down further.
4. Place the stopper back in the opening and reattach the rod, nut and screw, if necessary, or simply turn the stopper clockwise to lock it back in place.
5. Remove the container and turn on the tap. If the drain is still clogged, the next step is to use a plunger.

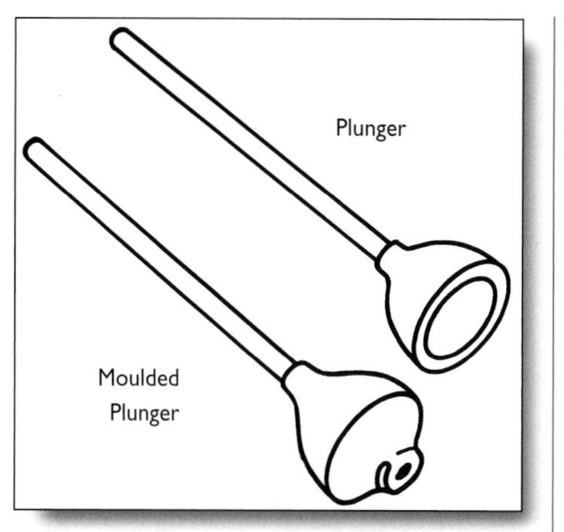

Use a plunger

There are two types of plungers—moulded and plain suction cup. Use a moulded plunger to clear a drain.

1. Remove the stopper or strainer.
2. Place the plunger over the drain. It works better if you plug the overflow drain (the hole in your sink or tub just under the taps or at front of the sink) with a cloth or piece of duct tape. This increases the suction effect of the plunger.
3. Turn on the water until there is about 50 mm (2 in.) of water in the sink or tub.
4. Move the plunger up and down several times. Be patient; it may take repeated attempts to clear the clog. If you are successful, the water in the basin will drain away once the plunger is removed. When this happens, turn on the tap water.
5. If the water drains properly, replace the stopper or strainers. If the drain is still clogged, try cleaning out the trap.

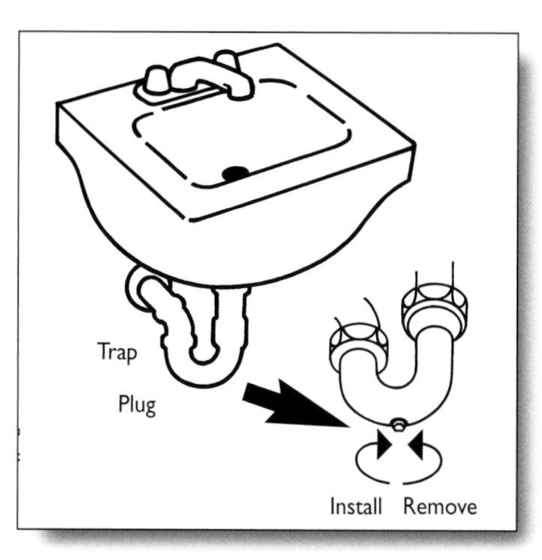

Cleaning out the trap

Trap with a plug

1. Place a container under the trap.
2. If the trap has a plug, remove it with a wrench or groove-joint pliers by turning it counter-clockwise. Let all the water drain out.
3. Cut the hook end off a wire coat hanger. Bend one end of the hanger into a small hook.
4. Push the hook into the trap, push and pull with it until the clog is cleared. Clean the trap opening with a brush or rag.
5. Reinstall the plug and turn on the water. Check for proper drainage.

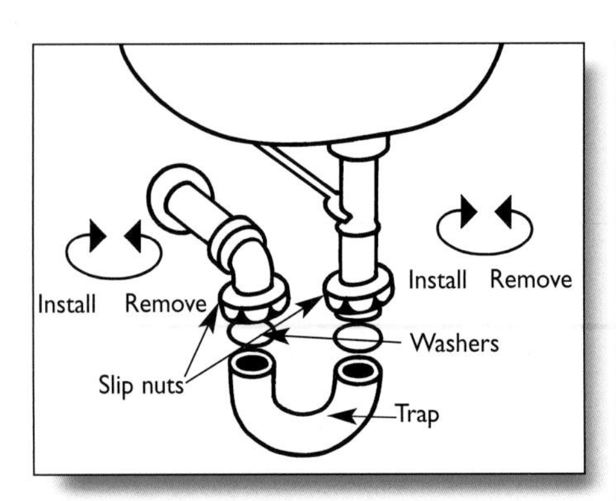

Trap without a plug

1. Loosen the two slip nuts with a wrench or groove-joint pliers. Note: Protect chrome pipes and fixtures with a cloth or with layers of tape before using a wrench on them.
2. Remove the trap and clear out any obstruction with a piece of wire. Clean the trap with a brush or rag.
3. Check to see that the two washers that fit between the trap and the nuts are not damaged. If they are, take them to your hardware store and buy matching replacements.
4. Reattach the trap by holding it firmly in position and tightening the two slip nuts.
5. Turn on the tap and check for leaks and proper drainage.

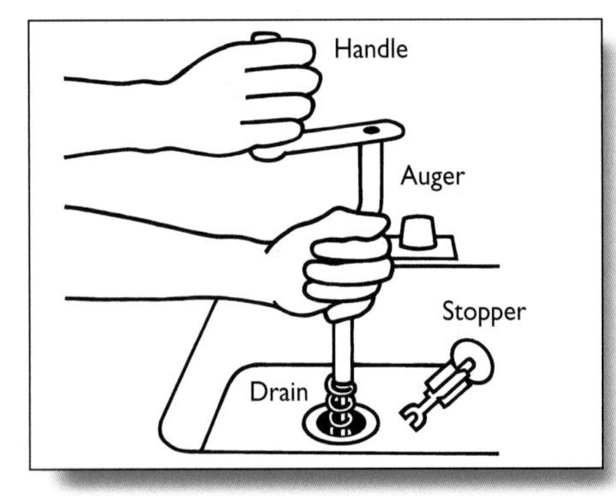

Use a drain auger

Through the sink opening (tailpiece)

1. Remove the stopper or strainer.
2. Push the auger into the sink drain opening until it stops. When you feel resistance, start turning the handle in one direction only. Exert some pressure, but do not force the auger.
3. Exert pressure and turn the auger handle so it moves further down the drain. Continue doing this until the auger moves freely in the pipe. Free movement means that you removed the clog.
4. Remove the auger and clean it. Replace the stopper or strainer and pour hot water into the sink. Check to see if the drain is working properly.

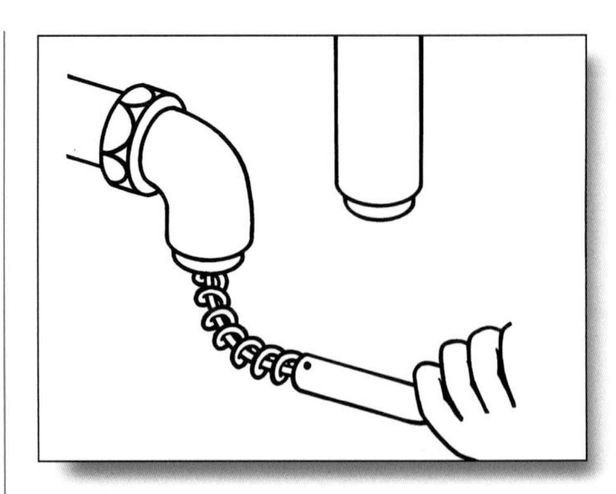

Through the sink trap

1. Remove the plug or trap.
2. Push the drain auger into the pipe until it stops. Exert moderate pressure and turn the handle in one direction only. The auger should move further into the pipe.
3. Continue until the auger moves freely.
4. Remove the auger and clean it. Replace the plug or trap and pour hot water into the sink.
5. Check to see that the drain is working.

Note: In the unlikely event that your drain is still not clear after trying all these methods, call a plumber.

Toilet Repairs

Diagnose toilet problems	
Problem	**Solutions**
Overflowing toilet	1. Use a plunger to clear the toilet. 2. If plunger doesn't work, use a toilet auger. If auger doesn't work, call a plumber.
Running toilet Water is more than 25 mm (1 in.) below top of overflow tube	1. Repair flush ball. Or 2. Clean the valve seat. Or 3. Adjust float ball guide.
Water doesn't shut off	1. Repair or replace the float ball. 2. Repair or replace inlet valve.
Toilet flushing Refill tube fallen out of overflow tube	1. Reattach refill tube so it stays inside overflow tube.
Loose handle	1. Tighten handle nut.
Loose or unscrewed lift chain	1. Tighten flush ball or replace. 2. Shorten chain so that the flush ball or flapper seals tightly.
Condensation on toilet tank	1. Insulate tank 2. Replace tank with an insulated tank
Leaking toilet seal	1. Replace toilet seal.

You can easily fix minor toilet problems. Most flush toilets are similar. Some toilets may have parts that look different than the one pictured, but all toilets work and are generally repaired in the same way. Newer toilets usually have a flapper instead of a tank ball to close the tank valve.

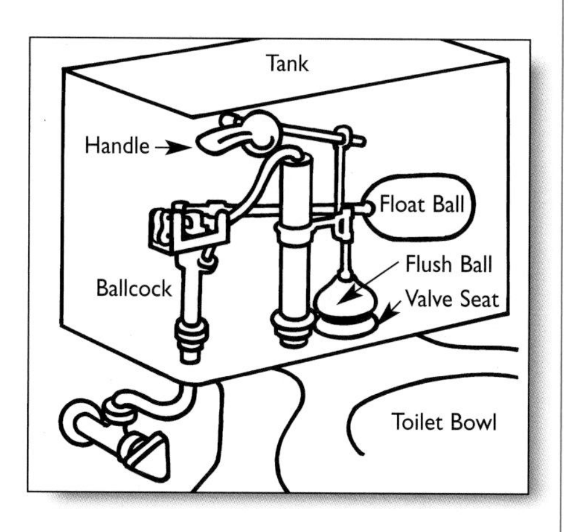

Before trying to repair a toilet, learn how the system works. The mechanism inside the tank is designed to produce enough water to flush the bowl completely. It works like this:

1. Depressing the handle causes the flapper (or flush ball) to rise, releasing water from the tank through the valve opening into the bowl.
2. The float ball drops with the level of the water in the tank.
3. As the tank empties, the flapper (or flush ball) sinks slowly back into place on the valve, shutting off the water flow to the bowl.
4. As the float ball drops, it lifts the inlet valve in the ballcock assembly unit that controls the flow of water into the tank.
5. Fresh water flows through the ballcock assembly into the tank. This causes the float ball to rise with the water level. As it rises, it depresses the inlet valve and shuts off the water flow as the tank fills.

Clear an overflowing toilet

Overflowing toilets are generally caused by something that is caught in the toilet trap. You have to try to dislodge this object, either with a plunger or a toilet auger.

Note: The only things that should be flushed down the toilet are human waste and toilet paper.

Skill level rating: 2 - Handy homeowner

Materials: water

Tools: screwdriver, plunger, wrench, groove-joint pliers, needle-nose pliers, funnel, drain auger

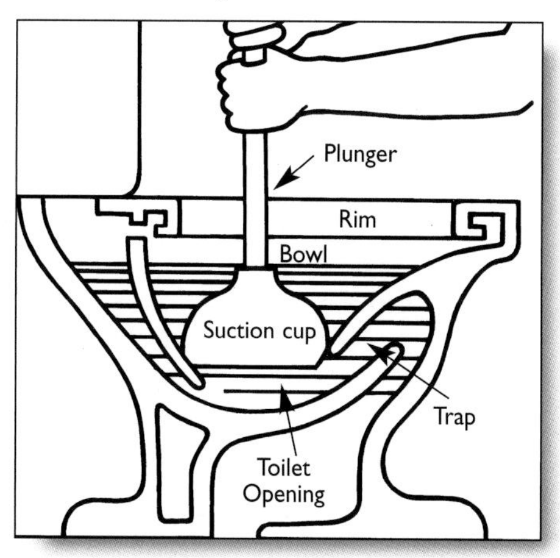

Use a plunger

1. If necessary, pour water into the toilet bowl until the bowl is half full. Do not flush the toilet. This action may release too much water and overflow the bowl.
2. Place the suction cup of the plunger over the toilet drain opening. A moulded plunger usually works best in toilets.
3. Move the plunger up and down until the water drains from the toilet upon removal of the plunger. Clean out the drain by flushing the toilet several times. If the toilet does not clear after using the plunger, use an auger.

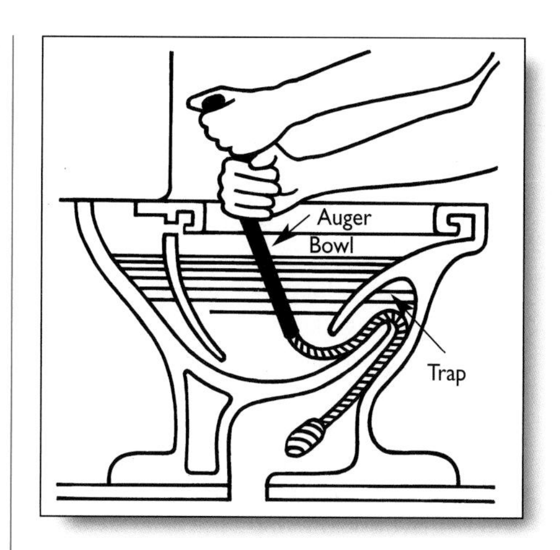

Use a toilet auger

Toilet augers are thicker than drain augers and have a long handle that protects the bowl from scratches.

Push the end of the auger into the toilet bowl drain opening.

1. Turn the handle and apply pressure until the auger is down as far as it will go. This action should clear the clog.
2. Remove the auger and pour a pail of water into the toilet bowl. If it drains away freely, the clog is gone. Flush the toilet several times to clear the drain completely. Clean the auger and put it away.

Note: If this method does not work, the clog may be in the main drain. In this case, call a plumber.

Fix running toilets

To find out where the problem is located, first remove the top of the tank and place it where it will not be damaged. Check the water level in the tank. If it is more than 25 mm (1 in.) below the top the overflow tube, the flapper (or flush ball) needs correcting. You will either have to repair it, clean the valve seat, or adjust the float ball guide.

If water is running into the top of the overflow tube, the float ball needs correcting. You will have to repair or replace it. If the problem is neither the flush ball nor the float ball, the inlet valve may be causing the problem. Replace the tank top when you are finished with your repairs.

Skill level rating: 3 - Skilled homeowner

Materials: will depend on the problem

Tools: coat hanger

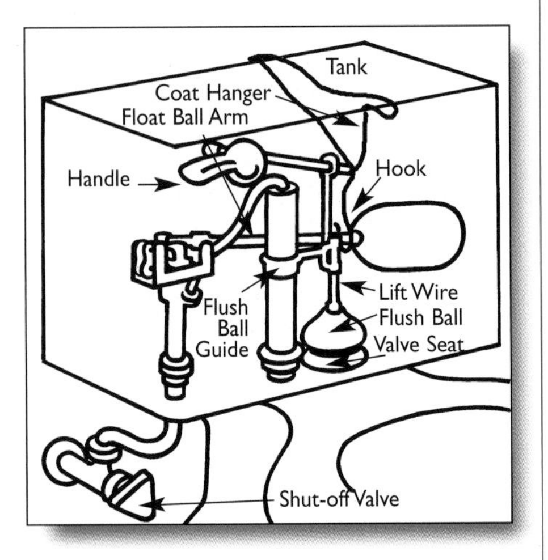

Repair the flush ball

1. Turn off the shut-off valve. If you do not have a shut-off valve, bend a coat hanger over the top of the tank. Place the hook around the float ball arm and bend the hook to keep the arm in its highest position. Unless there is something wrong with the inlet valve, the hook will keep the tank from filling while you fix the toilet.
2. Note: If there is something wrong with the inlet valve, you may have to shut off the main water supply while you first repair the flush ball.
3. Flush the toilet to empty the tank.
4. Check the flush ball to see that the lift chain is firmly attached, and whether there are signs of damage or wear. If the chain is loose, damaged or worn, replace the flush ball.
5. If the flush ball seems okay, the ball seat may need cleaning or the flush ball guide may need repair.
6. Remove the flush ball by holding the lift chain up and turning the ball clockwise.
7. Install a new flush ball of the same size and type on the lift chain by turning it counter-clockwise. Alternatively, you can replace the flush ball with a flapper.
8. To replace the flush ball with a flapper, remove the flush ball, lift chain and guide rod. Slide the collar for the flapper over and down to the base of the overflow tube. Position the new flapper over the valve seat.

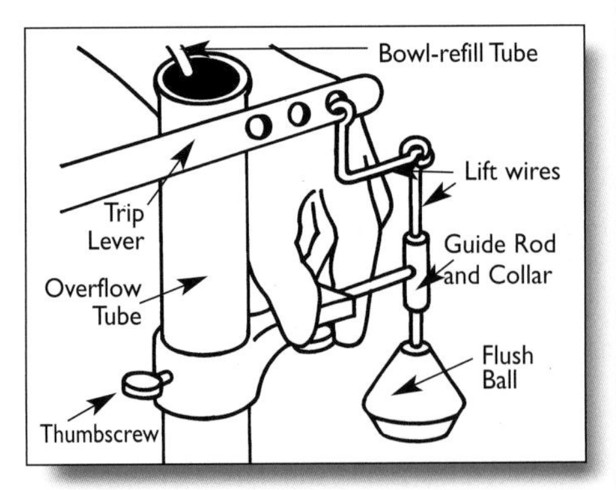

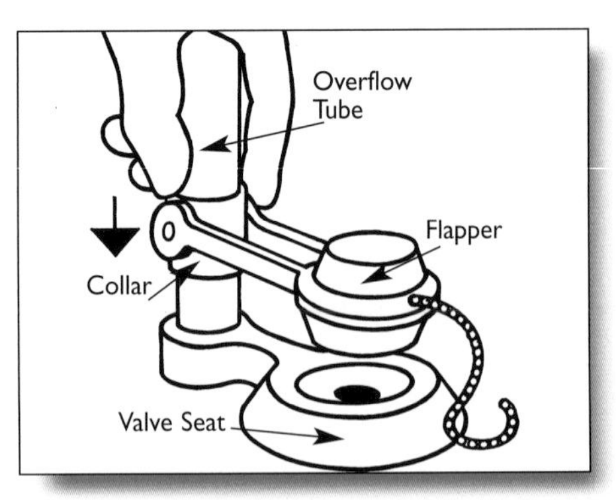

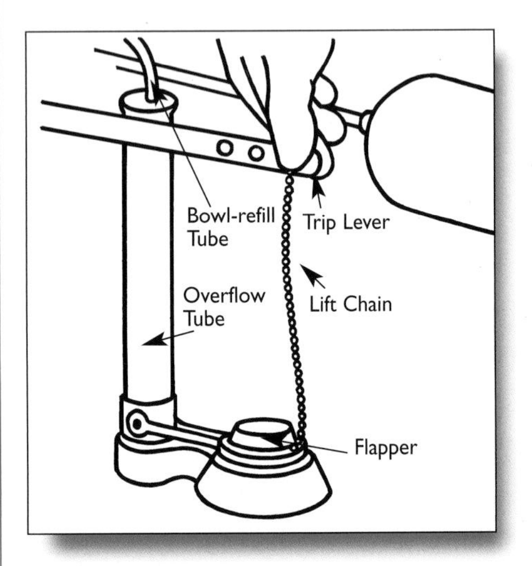

9. Adjust the lift chain and attach to the lift arm. There should be about 12 mm (1/2 in.) slack.
10. Depress and release the handle several times. Check to be sure the new flush ball or flapper sits evenly on the ball seat. Turn on the water or remove the coat hanger.
11. Check that the toilet is shutting off properly and that the water level rises to about 25 mm (1 in.) from the top of the overflow pipe.

Clean the valve seat

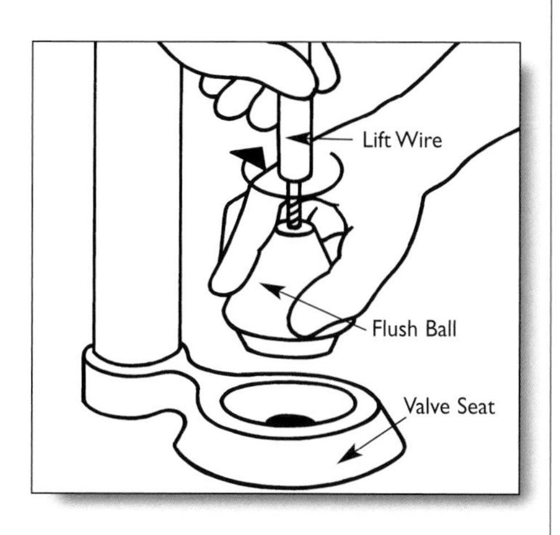

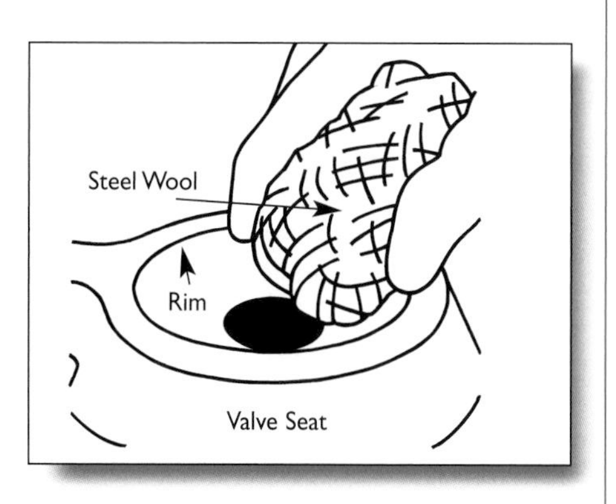

1. Shut off the water and flush the toilet.
2. Raise the flapper or flush ball and clean the rim and inner surface of the valve seat with steel wool.
3. Turn on the water.

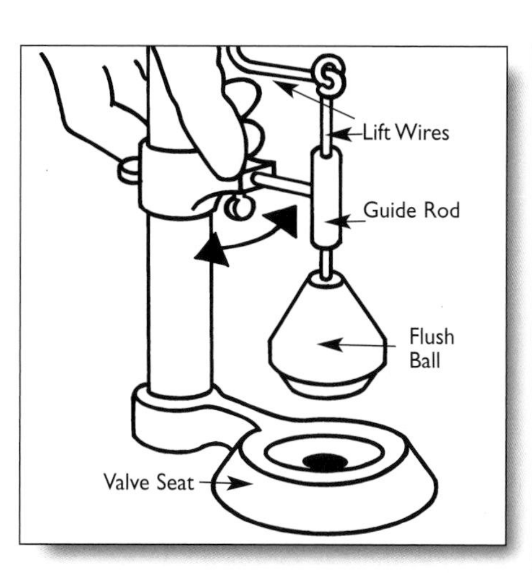

Adjust the flush ball guide

1. Turn off the water to the toilet.
2. Loosen the screw in the guide.
3. Move the guide left and right, and up and down until the flush ball sits evenly on the ball seat. Hold the guide in this position and tighten the screw.
4. Turn on the water to the toilet.
5. Flush the toilet and check to see that it flushes correctly. Some toilets have a flapper operated by a chain instead of a float ball and guide. To adjust, change the length of the chain until the flapper sits evenly on the ball seat.

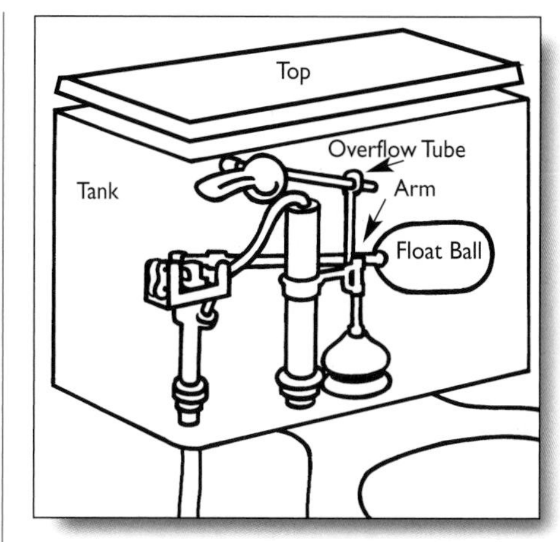

Repair or replace the float ball

1. Gently lift the float ball arm. If the water does not shut off, the inlet valve needs repair. If the water does shut off, the problem is in the float ball arm or the float ball itself.
2. Observe the float and the arm during a flush. If either touch the sides of the tank, bend the arm enough to correct the problem.
3. If the problem continues, remove the float ball by turning it counter-clockwise. If it contains water, replace it.
4. Attach a new float ball to the arm.
5. Adjust the arm until the new float ball is about 12 mm (1/2 in.) lower than the overflow tube. Flush the toilet and check whether the water shuts off when it is about 25 mm (1 in.) below the overflow tube.

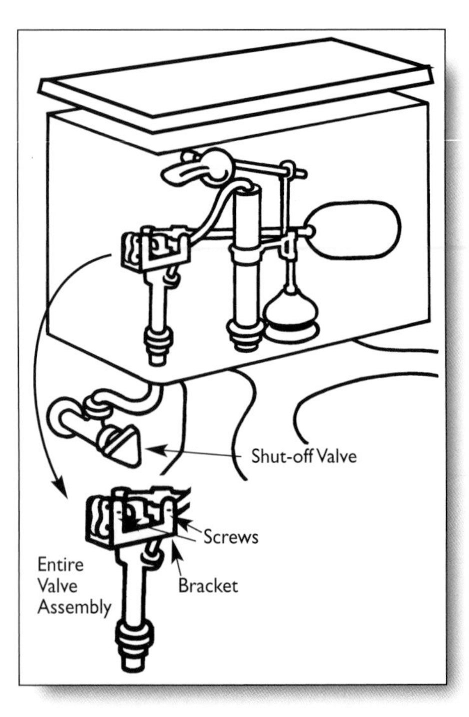

Fix the inlet valve

1. Turn off the water to the toilet.
2. Remove the screw in the linkage arm and lift out the inlet valve.
3. Check the washers for damage.
4. Install new washers of the same size and type and place the inlet valve on the linkage arm. Place the arm in position on the bracket and secure it with the screws.
5. Turn on the water to the toilet.
6. Flush the toilet to see if the inlet valve is working properly. Sealed inlet valves must be replaced.

If you are still having problems, replace the entire assembly with a modern inlet-float assembly. Your hardware store can give you instructions, or you can have a plumber install it.

Fix toilet flushing problems

If the toilet will not flush at all or flushes very slowly, locate the problem and repair the lift or connecting wires, or the chain.

Skill level rating: 2 - Handy homeowner

Materials: none

Tools: none

1. Check to see whether the refill tube has fallen out of the overflow tube. If it has, reattach the refill tube so it stays inside the overflow tube.
2. Check the handle to see if it is loose. If it is, hold the handle and tighten the nut on the inside of the tank with a wrench, but not too much because it can crack the tank.
3. Check the lift wire to see if it is loose or has come unscrewed from the flush ball. Tighten the flush ball onto the wire or replace it.
4. If your toilet has a flapper supported by a chain, disconnect the chain and shorten or lengthen it until it reaches from the arm to the flapper with about 12 mm (1/2 in.) of slack. Re-connect the chain to the trip arm.

Insulate or replace toilet tanks

When warm humid air comes in contact with a cold surface, the air becomes cooler and water vapour in it condenses into drops of water. This situation often occurs in a bathroom where the air is warm and humid and the toilet tank is cold. To minimize condensation on an uninsulated toilet tank, raise the surface temperature of the tank by insulating it or lower the relative humidity in the bathroom through ventilation or dehumidification. Another option is to replace the toilet tank with a new insulated tank. If the tank is already insulated, but still prone to condensation, you will need to lower the relative humidity in the bathroom.

Use the bathroom fan regularly to exhaust warm, moist air. Use a portable dehumidifier in the bathroom in hot, humid weather, if needed.

Skill level rating: 2 - Handy homeowner

Materials: toilet tank insulation kit or 12.5 mm (1/2 in.) polystyrene insulation, silicone sealant

Tools: caulking gun, utility knife, rag

1. Close the shut-off valve on the toilet water supply line.
2. Flush the toilet twice to empty the tank.
3. Wipe or sponge inside of the empty tank to dry it well.
4. Line the tank sides and bottom with polystyrene from a kit or cut pieces of polystyrene to fit.
5. Seal the insulation in place with silicone sealant.
6. Allow the sealant to dry for 24 hours.
7. Open the shut-off valve to refill the tank.

Replace wax toilet seal

Water on the floor around the toilet or a sewer gas smell may indicate that the seal at the base of the toilet has failed. Sometimes a toilet may not be firmly seated due to an uneven floor, loose flange where the toilet connects to the drain or loose bolts holding the toilet to the flange. Eventually this movement can cause the seal between the toilet and the floor flange to fail.

You can often fix the problem and reinstall the toilet. However, this problem may also be considered an opportunity to replace an older toilet that has an uninsulated tank with a new low-flush water usage toilet, 6 L/flush (1.3 gal.) with an insulated tank. The new toilet will use much less water and the insulated tank is less prone to condensation. These points are good to bring up with a landlord or property manager if they must make the decision.

Skill level rating: 2 - Handy homeowner, lifting the toilet may require two people

Materials: wax seal, toilet bolts, silicone caulking

Tools: wrenches, level, sponge, screwdriver, pail, putty knife, caulking gun

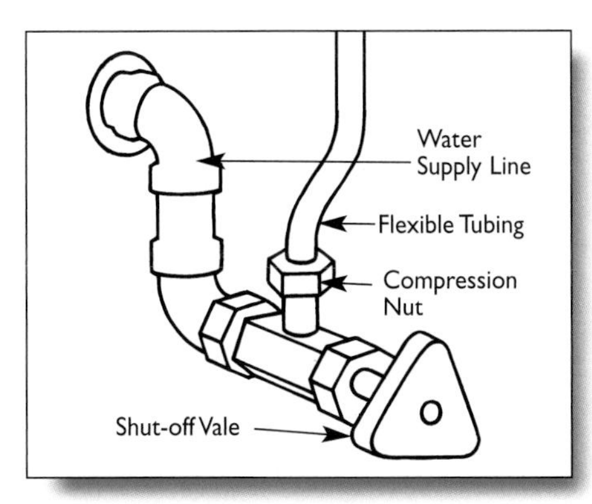

1. Turn off the shut-off valve on the water supply line to the toilet.
2. Flush the toilet twice.
3. Remove the lid on the toilet tank and set it aside carefully. Note: porcelain toilets are easily damaged.
4. Use a sponge to remove the remaining water from the tank and bowl. You may not be able to get all the water out.
5. Disconnect the water supply line at the tank.

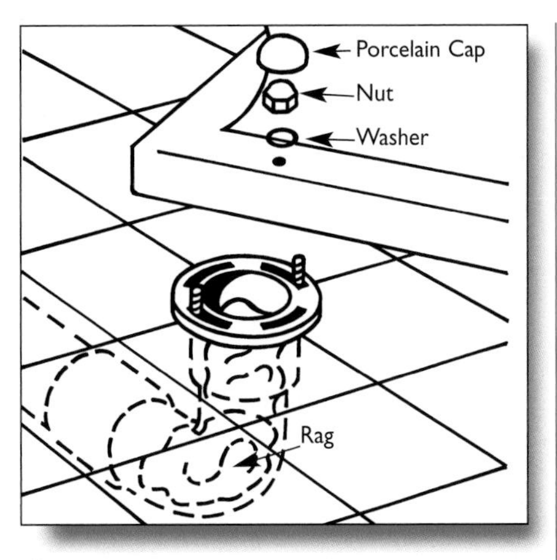

6. Disconnect the tank from the bowl by removing the tank bolts, if necessary.
7. Pry off or unscrew the caps at the base of the bowl that cover the toilet bolts.
8. Remove the nuts and washers from the toilet bolts.
9. Rock the toilet slightly to break the seal between the toilet and the flange.
10. Pull the toilet straight up. Try to catch any leftover water in a pail.
11. Temporarily plug the toilet drainpipe in the floor with a rag to prevent any debris from falling in or sewer gas from coming out.

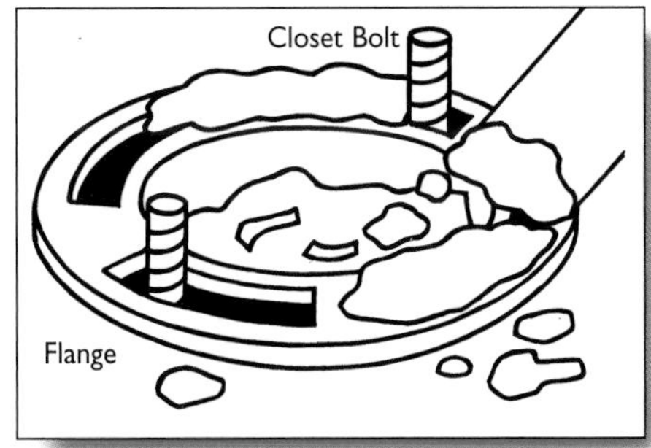

12. Scrape the floor flange for the toilet to remove any old wax or debris. Ensure that the flange is secure and at least 6 mm (1/4 in.) above the floor. If the flange is too low, you may be able to use a special gasket with a plastic sleeve that fits into the flange. If the flange is cracked, it will have to be replaced. This may require professional help.
13. Slide the old bolts out of the slots in the flange and replace with new bolts, if necessary. Align the bolts in the previous position and parallel to the wall.

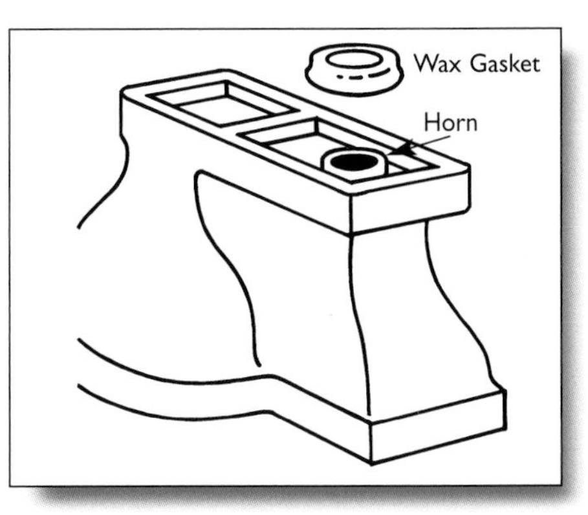

14. Carefully turn the toilet bowl over, remove the old wax seal, clean the discharge opening and press the new wax gasket in place.
15. Remove the temporary rag in the floor drain pipe.

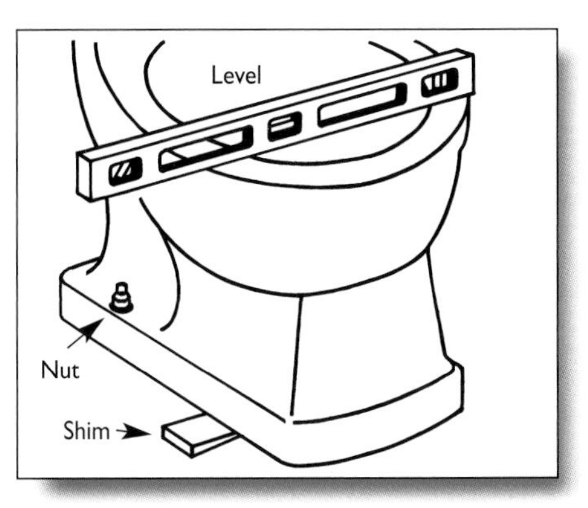

16. Turn the bowl upright and lower it carefully in place over the bolts and flange. Press and jiggle slightly until the toilet base is firmly seated on the floor. Replace the nuts and washers finger-tight on the bolts.
17. Ensure that the tank (reattach if necessary) is parallel to the wall. Carefully tighten the toilet bolts alternately. Note: over-tightening may crack the porcelain bowl. Snap the plastic caps back onto the toilet flange nuts and washers.
18. Re-connect the water supply line. Flush the toilet several times. Replace the tank lid. Check for leaks at the base of the tank and base of the toilet a few times over the next 24 hours.

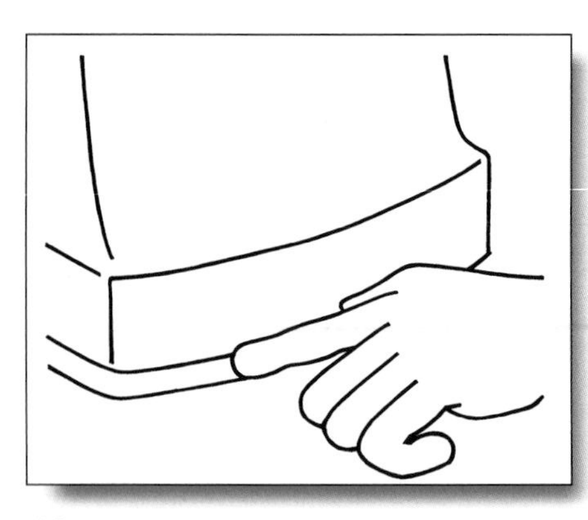

19. Caulk around the base of the toilet with mold-resistant silicone to ensure that any water on the bathroom floor doesn't get under the toilet where it may damage the floor.

Repair leaking shower stalls and tub surrounds

Shower stalls or tub surrounds may have finished walls of ceramic tile, fibre-reinforced plastic or acrylic resin. Water that gets behind any of these finished surfaces may cause damage to the supporting wall surface, framing or insulation and allow mold to grow. Water may leak behind the finish because of damage to the finished wall surface, deteriorated caulking or plumbing leaks.

Skill level rating: 3 - Skilled homeowner

Materials: mold resistant silicone caulking

Tools: caulking gun, putty knife, rags, detergent, water, rubber gloves, pail, masking tape

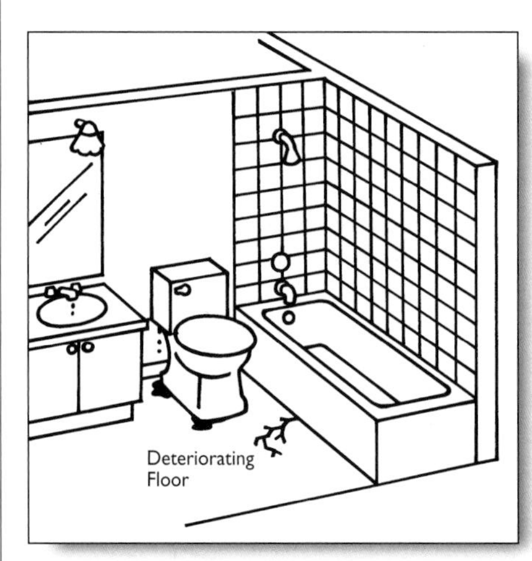

Assess and repair the tub surround or shower enclosure

1. Gently push against the walls above the tub. If the wall is soft or flexible, there may be damage or mold growth in the underlying drywall, insulation or framing. This will likely become a major renovation where professional help is required. Have a professional investigate further.
2. In particular, check around any window that may be above the tub. Because these windows are often the coldest surface in a warm, moist room, condensation may occur on them. If condensation is severe, water may run down the wall behind the tub surround. If condensation is a problem, clean up any mold (if there is any) as described in *Foundations and Basements* and consider better bathroom ventilation (see *Ventilation*).
3. Look for mold growth on any surfaces, particularly in the bead of caulking between the tub and tub surround. Clean up small areas of mold according to the mold cleanup instructions in *Foundations and Basements*.
4. Look for missing caulking between the tub and tub surround or around penetrations in ceramic tile walls such as soap dishes and faucets. If necessary, remove damaged or moldy caulking by scraping gently with a putty knife. When re-caulking, use two rows of masking tape to define the edges of the caulking bead. Cut the end of the caulking tube at a 45 degree angle and far enough back to produce a bead that will fill the joint. Apply kitchen and bathroom silicone caulking. Using a wet finger, quickly shape the bead. Strip off the masking tape to leave the bead of caulking with uniform edges.
5. Look for any damaged surfaces such as loose acrylic resin panels or damaged ceramic tiles. Make sure that the back of any loose acrylic resin panel is dry and that the wall behind the panel is in good condition before fastening the panel in place using a compatible adhesive (refer to the installation instructions or the caulking label to see if the adhesive can be used with the type of panel you are repairing). Caulk as noted above.
6. If you suspect water leaking behind fixtures, remove the wall cover (escutcheon) plates around the showerhead and taps. Unscrew the tub faucet. Each of these should have an upside down horseshoe shape of caulking behind the plate that prevents water from leaking in the top or sides but allows water to drain out the bottom. Scrape and remove old caulking, as required. Re-caulk with a kitchen- and bath-type silicone caulking before replacing wall cover plates.

Fix leaks in ABS drain pipes

Even slight water leaks that go unchecked can cause serious mold and rot problems. All water leaks should be repaired promptly. Investigate any suspicious water stains or slight drips before they cause damage. Metal drain fittings from sinks fit into ABS plastic drainpipes through connections that have tapered plastic compression washers. As the fitting is tightened (by hand), the washer seals around the metal pipe. Below that are traps that should always hold enough water to prevent sewer gases from coming back up through the drains. Many traps have a small clean-out port on the bottom with a screw type plug and washer. All plumbing drain lines, other than traps, should be sloped down appropriately and supported sufficiently to prevent sags in the pipe. All water must drain freely and all solids must be carried out.

Occasionally ABS plastic pipe joints may leak because the connection was never properly cemented.

Skill level rating: 3 - Skilled homeowner or may be

Skill level rating: 4 - Qualified tradesperson/contractor. Check local requirements.

Materials: required ABS pipe, fittings and straight connectors, ABS solvent cement, compression washers, pipe strapping, wood screws

Tools: hacksaw, mitre box, tape measure, pencil or marker, utility knife, cement brush

1. Examine all compression fittings and trap clean-out plugs. If there are any signs of leaks, tighten gently. If leaks persist, disassemble the fitting; clean thoroughly; install a new compression washer; re-assemble and tighten gently. Check for leaks.
2. Examine accessible pipes (usually in the basement or crawl space) for leaks or sags.
3. If there is a leak at a fitting, it is usually best to cut the fitting out and replace it.
4. Ensure that the drain will not be in use during repair.
5. Cut the pipe a few inches to each side of the fitting.
6. Use two straight connectors (sleeves), two short lengths of pipe and a new fitting to replace the old one.
7. Carefully cut the short lengths of pipe to the right lengths.

8. With a utility knife, smooth the inside of the cut and bevel the outside edge.

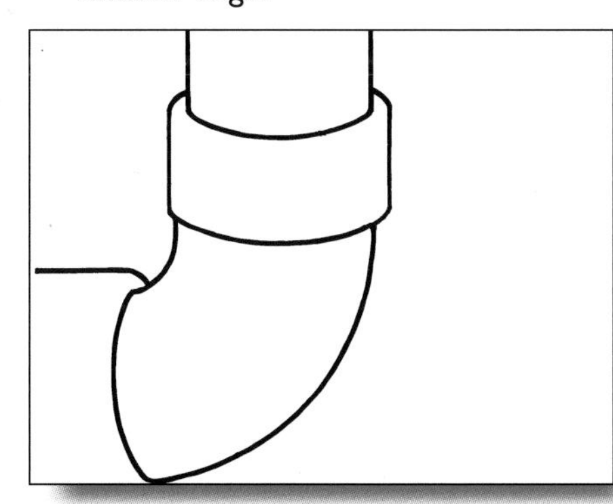

9. Dry fit all pieces together before using any cement.

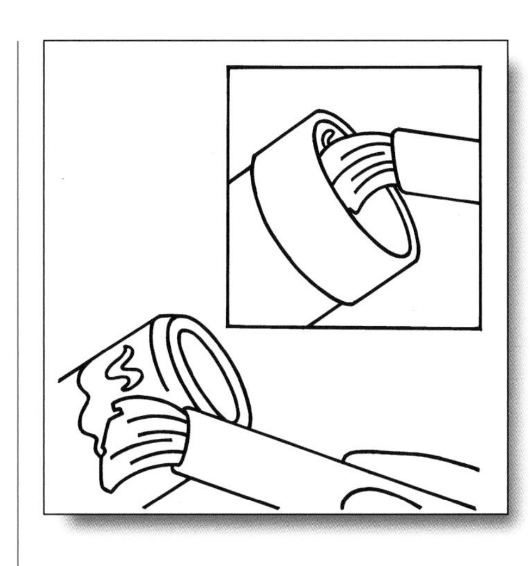

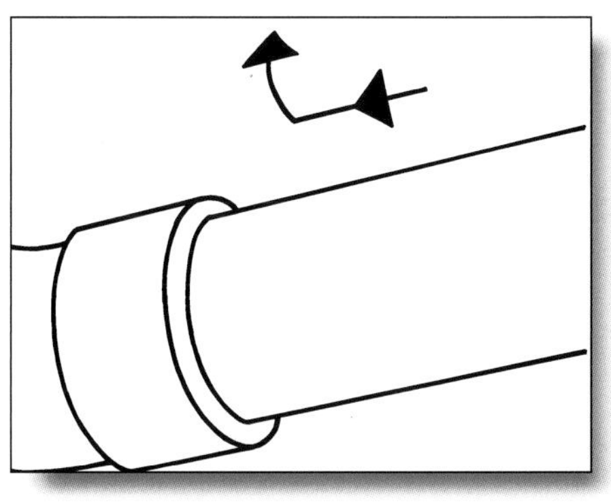

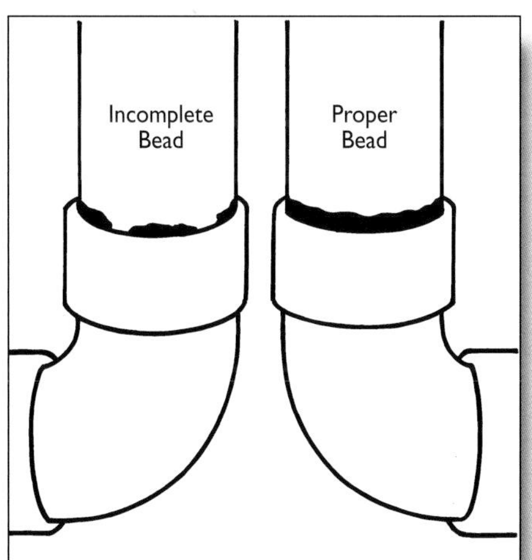

10. Do each connection in a sequence that will allow you to fit the final connection in position. Apply a liberal amount of solvent cement to both mating surfaces. Slide the pipe fully into the fitting and twist it into final alignment. Remember that cement sets in less than a minute so work quickly. Look for a small continuous bead of cement all the way around the joint to ensure a complete seal. Hold the two pieces in the proper alignment for at least 30 seconds.

11. After a few minutes, repeat this procedure on the next connection until the complete drain line has been successfully re-assembled.
12. Add additional straps to support any sags in long horizontal runs of pipe.
13. Wait 12 hours before running water through the pipe.
14. Check for leaks.

Secure noisy pipes

Plumbing noises may be caused by water supplies or by drainpipes. Water hammer is a banging noise when a faucet is turned off. This occurs because water is travelling under pressure in a supply pipe and it has suddenly been stopped. The sudden stoppage shakes the pipe. Securing the pipe may stop the noise or you may need to install a water hammer arrestor, or air chamber, at the fixture. The chamber provides a cushion of air for the water pressure. Drainpipes may also bang if they are not sufficiently secured.

Gurgling noises or poor drainage may indicate an obstructed, poorly installed or missing vent system. Check for obvious obstructions where the vent exits the roof. If there are none, have a plumber inspect the system.

Skill level rating: 2 - Handy homeowner

Materials: Copper pipe clamps and nails (for copper pipe); steel strapping and wood screws (for plastic pipe); any rubber cushioning material

Tools: hammer, screwdriver

1. Check for pipe hangers that are too loose or too tight. If pipes are accessible (usually in a basement or crawl space), gently try to jiggle them. If they rattle, you've found a place that needs better support.
2. Turn on the hot water. The heat causes cold pipes to expand. If you listen along the hot water supply line and hear a ticking sound, the pipe is secured too tightly and cannot expand along its length.

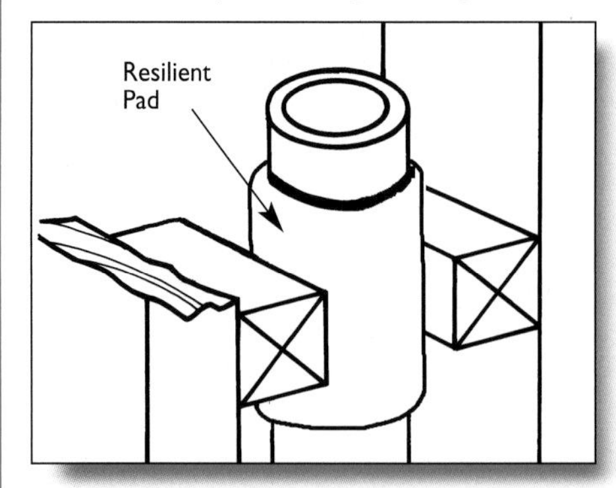

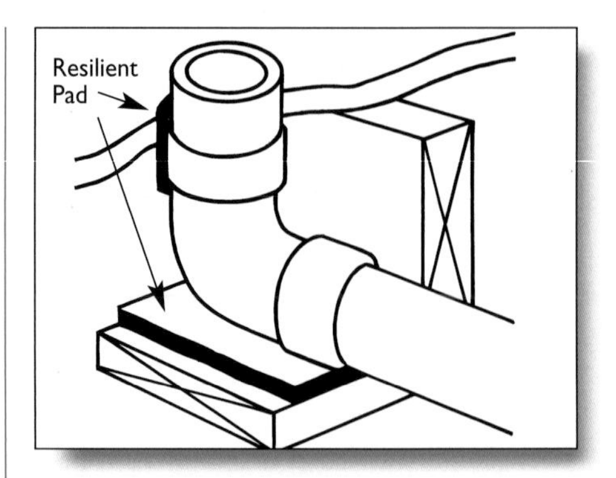

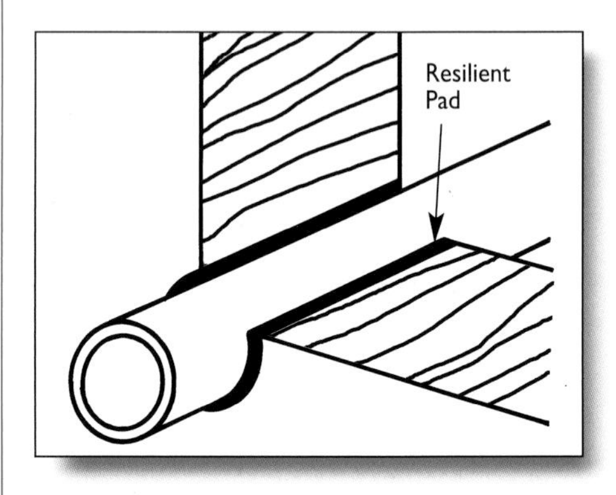

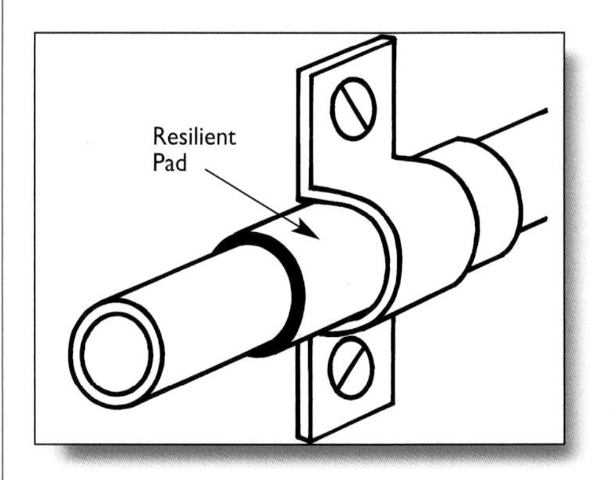

3. Adjust copper pipe straps as necessary ensuring that they are neither too loose nor too tight. Add a piece of rubber between the pipe and the strap for a cushion. Use only copper straps and nails with copper pipe to avoid corrosion caused by a reaction between dissimilar metals. Add more straps as required.
4. Adjust or add steel straps and rubber cushions to the drainpipes as required. Make sure there are no sags in long horizontal runs.

Thaw frozen pipes

If water lines freeze outside your home, call your local works department. Don't repair them yourself! If lines freeze inside your home, you can try to solve the problem yourself or call a plumber.

Tip

- If the pipes in your home freeze from time to time, insulate them or install a heater cable. Special pipe insulation and pipe heating cables are available from your local hardware or building supply store.

Skill level rating: 2 - Handy homeowner

Materials: epoxy paste, rubber pad, hose clamp, joint tape, cloths

Tools: heater cable or pad or vacuum cleaner or hair dryer or heat gun, rubber gloves

Thaw frozen lines inside

1. Turn off the water and try to find out exactly where the pipes are frozen. Look in places where the pipes are likely to get cold, such as in a crawlspace, an unheated cellar or under a mobile home behind the skirting.
2. Inspect visible areas of frozen lines for splits or cracks. If you find any, you can usually repair them temporarily until a plumber comes. Wrap tape around small leaks and use an epoxy paste on joint leaks. Make sure the pipes are dry first. Another option is to use a rubber pad, held in place by a hose clamp. Larger cracks can be temporarily sealed with a rubber or vinyl pad, held in place by a bolted pipe clamp.

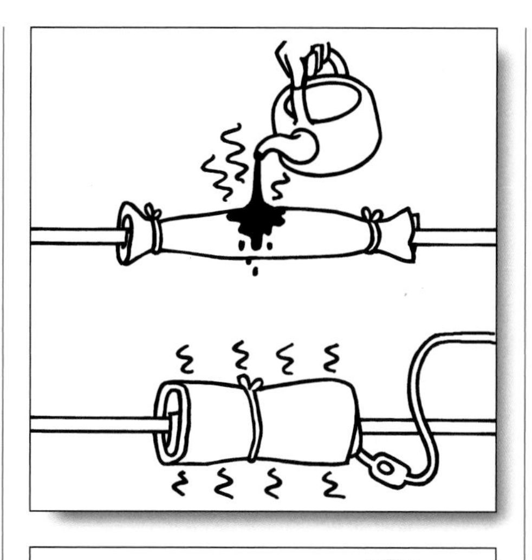

3. If no splits or cracks are visible, use gentle heat to thaw the frozen area. Here are some methods that you can use:
   a. Wrap cloths around the frozen pipe and wet them with hot water. OR
   b. Wrap an electric heater cable or heating pad around the frozen section. OR
   c. Blow air on the frozen section using your vacuum cleaner wand, attached to the exhaust of your vacuum cleaner. OR
   d. Blow air on the frozen section using a hair dryer or heat gun but do not hold it too close. Wear rubber gloves to prevent an electric shock.
4. Open all taps that are connected to the frozen lines so melting ice can run out of them. Always proceed with caution. Never heat pipes to a temperature higher than your hand can stand. Never use a torch. It makes intense, not gentle heat. Work from an open tap toward the frozen area to prevent steam from being trapped behind the ice and bursting the pipe. When the pipe begins to thaw, water will drip from the tap.

Repair problems with water pumps and pressure tank

A waterlogged tank or a faulty pressure switch may cause problems with a water pump that either will not start, or starts and stops too frequently. A pressure tank should be filled partly with air and partly with water. Some tanks have an air valve so that after the tank is drained down, the tank can be slightly pressurized (usually 15 psi or less) using any type of small air pump. Newer water tanks have a diaphragm that separates the air cushion from the water and is less prone to problems. Either way, the air compresses and provides water pressure in the supply lines without the pump turning on until pressure has diminished to a level set at the pressure switch.

A unique problem in Alberta is natural gas seeping into drilled water wells, which can also cause air blasts, in which case a degasifier may be required. In some cases, the problem is so extreme that when a lighter is turned on next to the faucet just after the air blasts occur, the flame will keep burning at the tap for a second or two. Call a plumber to fix this problem.

If, after following the instructions below, the tank seems to be operating properly but the pump still doesn't run well, the pressure switch may be defective. Consult a plumber.

Skill level rating: 3 - Skilled homeowner

Materials: none

Tools: hose or pail

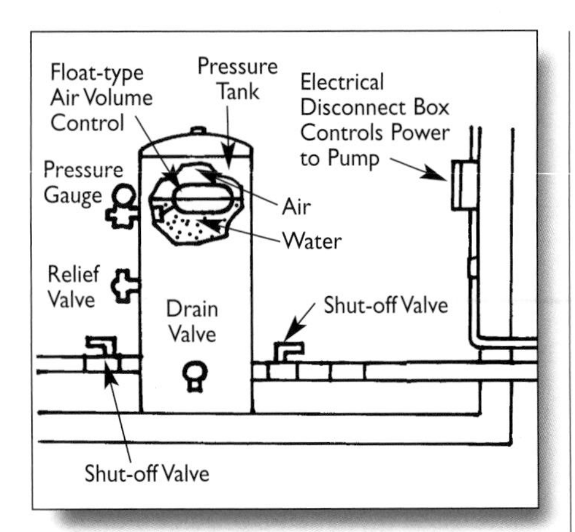

Fix a waterlogged tank

If there is no air pressure in the tank, it is referred to as being "waterlogged" and the pump will start and stop every time water is used.

1. Shut off power to the pump.
2. Attach a hose to the drain valve at the bottom of the tank.
3. Open the drain valve and a faucet in the house. Drain all the water out of the tank.
4. Turn off the faucet, close the drain valve and remove the hose.
5. Turn the power on to the pump.

Fix an airlock

If air blasts out of the faucets occasionally, the pressure tank may have an air lock.

1. Shut off power to the pump.
2. Attach a hose to the drain valve at the bottom of the tank.
3. Open the drain valve. Leave it open until there is no more pressure in the tank.
4. Close the drain valve and remove the hose.
5. Turn the power on to the pump.
6. If the tank has a diaphragm, the air lock may have been caused by a leak in the pipe between the well and the tank. Check all fittings. Consult a plumber if there are further problems.
7. Check for leaks.

Maintain hot water heaters

Water heating typically accounts for 20 per cent of the total energy used in a home, so it pays to keep your hot water tank in good repair. Hot water tanks can be purchased either from hardware or building supply stores or often from your local utility company. Rental programs are also usually available from your local hydro or gas company.

Always have a qualified person install and maintain the water heater. His expertise is especially important if you have a gas or oil water heater with a pilot light and burner. Gas and oil water heaters burn fuel when they heat the water. In any situation where burning or combustion occurs, it's important that the fuel is totally consumed for the system to operate safely and efficiently. When improper combustion occurs, gases are given off that can harm you. A qualified contractor or installer from the utility will ensure that your gas hot water tank is operating safely.

Electric water heaters have elements that sometimes burn out and need to be replaced. This happens more often in areas with hard water because the elements become coated with minerals, overheat and burn out. Call your local electric utility or plumbing contractor if your water heater is not operating properly.

Adjust temperature

Skill level rating: 2 - Handy homeowner

Materials: none

Tools: screwdriver

1. Adjust the water temperature to as low a point as practical for your household. Electric water heaters usually have two elements, one at the top and one at the bottom. Both of these elements have adjustable thermostats.
   Note: Keep the thermostat at 60°C (140°F). A lower setting can allow legionella bacteria to grow that can be a serious health risk.
2. To adjust the thermostats, remove the panels near the top and at the bottom of the tank.
3. Turn the small setscrew to a lower setting. The factory settings are usually at 65°C (about 145°F.). Your electric utility company can advise you about adjustments or repairs.

Gas and oil water heaters have one adjustment control, on the outside of the tank, at the bottom. Your gas or oil utility can advise you about adjustments or repairs.

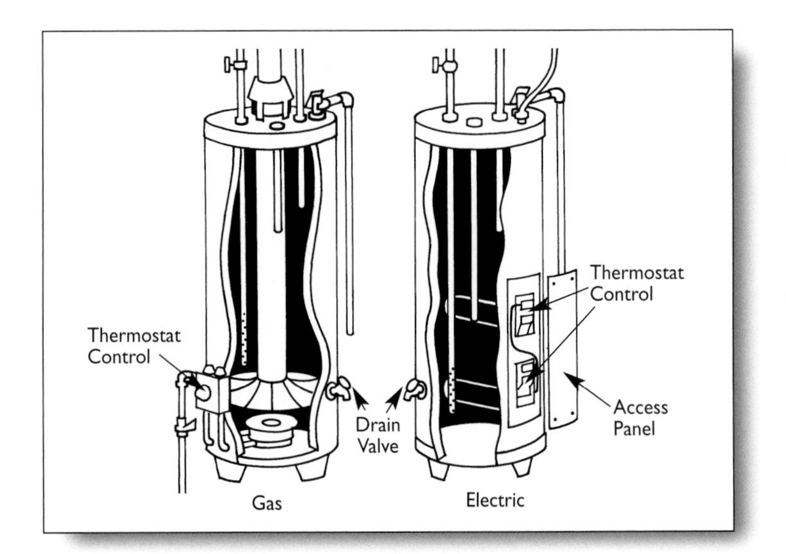

Prevent sediment accumulation

Sediment will build up in a hot water heater, especially in homes that have high mineral levels in the water. Draining part of the water that is stored in the tank will help to prevent sediment from accumulating and will extend the life of the tank. This should be done every six months.

CAUTION: The water coming from the tank is hot. Be extremely careful so no one gets burned.

Skill level rating: 2 - Handy homeowner

Materials: none

Tools: garden hose, bucket

1. Attach the hose to the water heater drain at the bottom of the tank.
2. Run the other end of the hose into a floor drain, to a sink or outdoors if an option.
3. Open the drain valve in the tank to let the water flow.
4. Allow the tank to drain until the water looks clear.
5. Close the drain valve. Remove the hose.
   Note: If the water doesn't run clear, you may need to shut off water supply and then empty and refill the water heater and repeat the flushing process.

Testing the temperature and pressure relief valve

Test the temperature and pressure relief valve on your hot water heater once a year. The relief valve prevents too much pressure building up in your tank. If not tested regularly, minerals in the water can prevent this valve from functioning properly.

Skill level rating: 2 - Handy homeowner

Skill level rating: 4 - Qualified tradesperson/contractor (plumber) if valve has to be replaced.

Materials: none

Tools: bucket

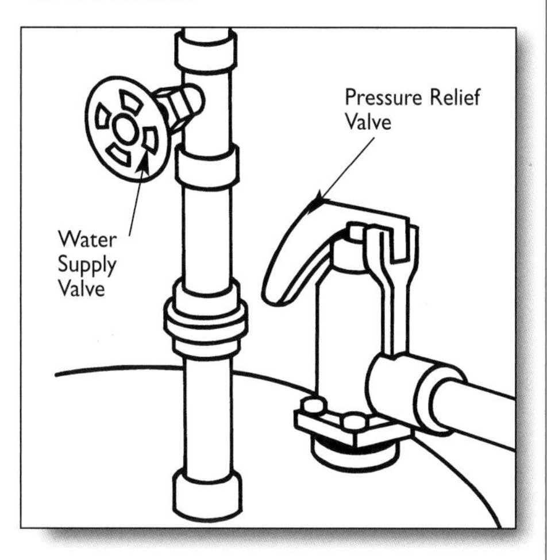

The relief valve is usually on the top or the side of the water heater and should have an overflow pipe attached.

CAUTION: Testing the relief valve may release steam or hot water that can cause burns.

1. Shut off electricity to the water heater. For an oil or gas water heater, turn down the thermostat so the heater does not start up.
2. Place the bucket under the overflow pipe.
3. Lift or depress the pressure relief valve handle. Water should drain out of the overflow pipe.
4. Turn electricity back on or adjust the thermostat to its original setting (oil or gas).

If water doesn't drain out or leaks after you've tested it, the valve is not working properly. Try opening and closing it a few times to get it to seal properly. If the valve continues to leak, call a plumber to replace the valve.

Replace the sacrificial anode

The sacrificial anode is a rod that is screwed into the top of the tank and prevents the tank from rusting. The rod is made of magnesium or aluminum, which is wrapped around a steel core wire. The rod needs to be replaced usually every five years to prolong the lifespan of the water heater. It should be replaced sooner if a 150 mm (6 in.) section of the steel wire is exposed or if the rod has a hard calcium coating.

Skill level rating: 3 - Skilled homeowner

Materials: replacement anode to fit the make and model of your hot water tank, if there isn't much overhead clearance, you may need to use a segmented rod so that you can get it into the tank

Tools: wrenches to fit the anode screw

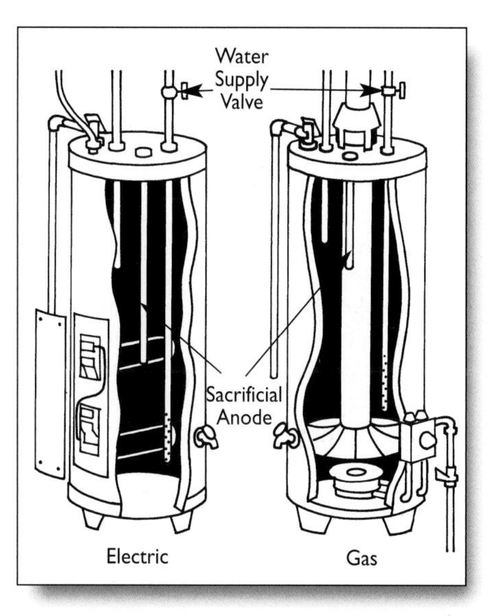

1. Shut off the water supply and electricity to the heater. For an oil or gas water heater, turn down the thermostat so the heater does not start up.
2. Unscrew the rod and remove it.
3. Insert the replacement rod and tighten it into the top of the tank.
4. Turn water and electricity back on or adjust thermostat to original setting (oil or gas).

Maintain septic system

Maintaining your septic tank properly will keep it operating better and help prevent expensive repair or replacement costs.

Skill level rating: 2 - Handy homeowner

Skill level rating: 5 - Specialist/Expert to pump the septic system

Materials: none

Tools: measuring stick

1. Have the tank inspected every year to measure sludge (settled solids) and scum levels (floating sewage). Tanks should be cleaned when the scum level is within 75 mm (3 in.) of the bottom of the outlet device or if the sludge depth exceeds 600 mm (24 in.). Tanks generally need to be pumped out every two to three years. Consult the yellow pages in your phone book to find a septic pumping service. A properly installed and maintained system lasts about 25 years.
2. Septic tanks use bacterial action to decompose the wastes. Use cleaning products that are biodegradable and do not harm the bacteria in your tank. Large amounts of bleach, lye, acids or disinfectants will kill the bacteria and stop your septic tank from functioning properly.
3. Try to keep your water consumption steady. Your septic system will be overloaded by a sudden increase in water.
4. Septic systems need to "breathe" in order to function well. Tree roots, pavement and driving cars on or near the system can stop your system from working well.

Open discharge septic systems are still widely used and meet (PSDS) **P**rivate **S**ewage **D**isposal **S**ystems codes in Alberta. Rocks should be piled up around the ejector line. A fence should always be installed around the ejector line and lagoon area to keep pets and children out of the main shoot-out area. The lagoons have proven to be a deadly "attraction" for children. It is important to educate children about the hazards.

1. Direct roof water, storm water, surface water and foundation drainage water away from the tank disposal field.
2. Put paper towels, newspaper, disposable diapers and sanitary napkins in the garbage, not into your toilet. These items often clog pipes and septic systems.
3. If you suspect a leak in your septic system, have it checked immediately. A leaky system can contaminate the ground water and make people sick.

Maintain water treatment equipment

Water softeners and particle filters are the most common types of water treatment equipment. The chemicals in water softeners need to be replaced periodically. Follow the manufacturer's instructions.

Simple particle filters are installed in a horizontal section of the water supply pipe. Filters must be changed periodically.

Skill level rating: 2 - Handy homeowner

Materials: replacement filter cartridge

Tools: rag

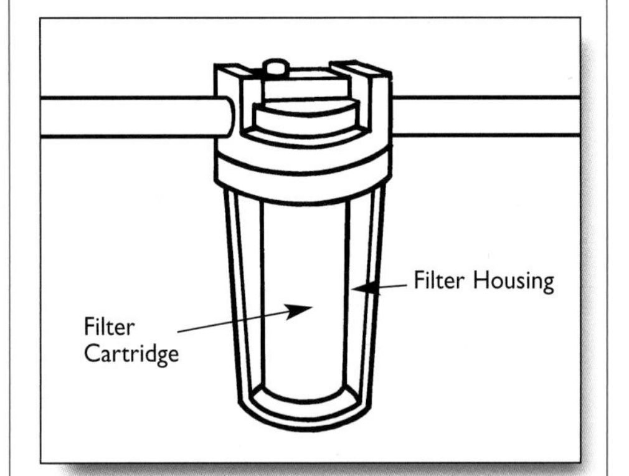

Change a filter cartridge

1. Most filter units have a built-in shut-off valve with the ON and OFF positions marked on the unit. Usually, turn a knob on the top of the unit 1/8 turn to the OFF position.
2. Grasp the filter housing with two hands and turn counter-clockwise about 1/8 turn to release the housing and filter cartridge from the unit.
3. Lower the filter housing from the unit. Be careful—it will be full of water.
4. Empty the water from the housing. Note which end of the filter is up. Discard the old filter. Wipe out the housing with a clean rag.
5. Insert a new filter into the housing with the correct end up.
6. Slide the filter housing up into place and secure by turning clockwise until it locks into position (usually 1/8 turn or less).
7. Turn the knob on the unit back to the ON position.

Maintain sump pumps

When the drainage tile around the outside of the house cannot disperse water from around a foundation, sump pits and pumps are used to relieve the water before it rises above the basement floor. Sometimes perimeter drainage tiles cannot be drained elsewhere and so they are routed into the sump pit where the water can be pumped out. In other cases, the sump pit just accepts water from weeping tile or a gravel drainage layer under the basement floor and drains to the sewer system or a dry well.

Sump pumps may be either submersible or pedestal types. When water in the pit reaches a certain height, a float-activated switch turns on the pump. When the float lowers far enough as the water subsides, the pump shuts off. Water is pumped out through a discharge line that dumps outside and away from the house.

Sump pump systems that are inspected and maintained regularly are less likely to fail during an emergency. Sump pumps fail for a number of reasons. Inlet pipes from perimeter drain tiles or from under the floor may become obstructed. Pump motors may burn out. Pump switches may fail to activate the pump. Power outages may eliminate power to the pump. Discharge lines may become disconnected; they may freeze outside; or they may dump water too close to the house so that it just runs back in. An unsealed or non-existent pump lid may allow moisture, mold spores or soil gasses into the house.

Skill level rating: 3 - Skilled homeowner

Materials: contractors sheathing tape, bucket or hose with water

Tools: hose, dust mask, goggles

1. Open the sump pit cover. The pit may contain mold. Wear goggles and dust mask as a basic safety precaution.
2. Inspect the pit and inlet lines. Remove any debris or obstructions. Check the inlet screen at the pump base to ensure that it is clear.
3. Check the breaker and electrical connection to ensure that the pump should be operational.

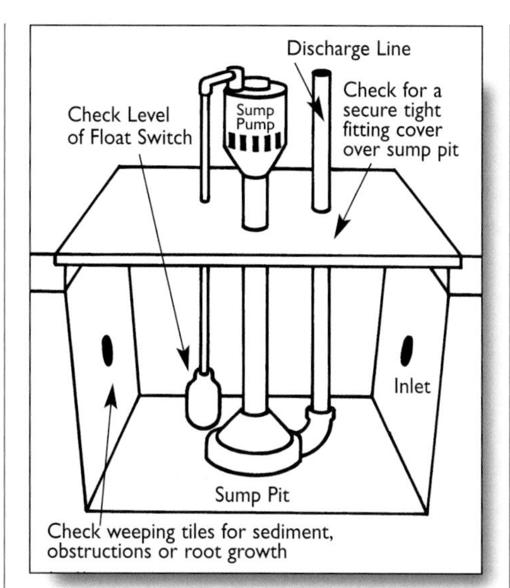

4. Use a hose or bucket to pour water into the pit. Make sure that the float switch starts and stops the pump at an appropriate water level. If the start and stop water levels are not appropriate, adjust the float level and try the test again.
5. If the pump does not work at all, re-check the electrical connection. If that seems fine, the pump switch or pump may need replacement.
6. Check the discharge line for leaks. Ensure that any hose clamps are tight and fittings are secure.
7. Check where the end of the discharge line empties outside the house. If the line is not sloped so that it drains quickly, it may freeze and be inoperable when needed most. Adjust slope as required. If the line terminates just outside the wall of the house, make sure that splash blocks direct water away from the foundation.
8. Replace the sump pit cover. Make sure that it fits tightly to prevent entry of moisture, mold spores and soil gases into the house. Tape slots or holes in the cover, as required. If necessary, make a new cover from plywood sealed in heavy polyethylene.
9. If power failure during emergencies is a concern, consider purchasing a sump pump system with battery back-up.

ELECTRICAL

Electrical power comes to your home from your hydro company. It flows through the meter, which records how much electricity you use. It then passes in to the main power control switch located near your fuse or circuit breaker box. From there, it travels to all electrical fixtures and outlets throughout your home.

The most common electrical problems are:

- blown fuses and tripped breakers
- broken wall outlets
- broken wall switches
- damaged cord plugs
- burned out light bulbs
- broken light fixtures

Maintenance includes:

- checking for any damaged switches, outlets, cords and plugs.

Prevention tips

CAUTION: Never forget that electricity can seriously injure or even kill you.

- Always follow the instructions that come with electrical appliances and never attempt difficult repairs.
- When changing or replacing fuses, switches and outlets always make sure the main switch is shut OFF. If possible, also lock the main switch OFF so you or someone else cannot accidentally turn it on. An easy way to jam the switch OFF so it can't be accidentally tripped to ON is to insert a screwdriver through the lock hole in the handle. If the ground is damp near the box, place dry boards over the damp areas and stand on them. Wear rubber-soled shoes to help protect you against an electric shock.
- Use common sense and always think about possible shock hazards before acting. Do not use appliances if they, you, or the area you are standing in, is damp or wet.
- Know where to find the main power switch and circuit box so you can act quickly in an emergency. Usually it is located **inside** your house, close to where the electrical meter is located (now usually **outside**).
- Do not use extension cords if you can avoid it. In a new home, extension cords should be unnecessary. Outlets are placed throughout new houses at convenient locations and should not be more than two metres (six feet) from where you want to place your appliance. Do not put more than two plugs into any one outlet. That's what the outlet is designed to hold. Do not take risks.
- Do not overload electric outlets. Heaters, electric kettles, toasters and older refrigerators use a lot of electricity. Stoves and dryers each need to have a heavy-duty, specialty outlet. In a new house, outlets specially designed for heavy loads are usually found in the kitchen, just above the kitchen counter.
- Only a qualified, licensed electrician should install extra electrical outlets. This point is very important. Note: Your electrical system was inspected when the house was built. If an electrical fire occurs, any improper changes or additions could result in loss of coverage by your insurance company. Also, in some locations, you may not be permitted to do electrical repairs. Check with your local utility company.
- A circuit tester is a handy diagnostic tool that allows you to check whether outlets are grounded and whether current is flowing. Testers are available at your local hardware store and may save you from a careless mistake.
- Plug grounded (three-prong) plugs only into grounded (three-prong) outlets. Do not alter the plug to fit into the outlet.
- Check to see whether the light bulb is screwed in properly or burned out before replacing the fixture.

Repair tips

- If you have a fuse box or an electrical stove, it's a good idea to keep a few spare fuses of various ratings around in case you have a problem when stores are closed. If the cartridge fuses in your switch box keep blowing, it's time to call an electrician.
- Choose a new fixture that is rated the same as the old one.

Special considerations

Healthy Housing™

- Use energy-efficient lighting such as compact fluorescent bulbs.
- If an appliance needs to be repaired, consider whether buying a new appliance may be a better, cost-effective solution. New appliances use less energy than older or re-conditioned units. They may cost more to buy, but will provide significant savings on electricity costs. When selecting a new appliance, compare the EnerGuide ratings to help you estimate the energy savings you will gain. Choose the model with the most favorable ratings to help reduce your household energy consumption.

Safety

- Check your appliances for labels or stickers indicating they are approved either by the Canadian Standards Association (CSA) or by Underwriters' Laboratories of Canada (ULC). Do not use unapproved appliances.
- Keep electrical appliances in good working order. Examine them for frayed cords, bare wires, and loose connections at plugs and outlets. Make repairs promptly.
- Do not use light bulbs that have a higher wattage than the rating of the electrical fixture.
- Protect children by installing childproof wall outlet covers on wall outlets.
- Ensure that ground fault circuit interrupter (GFCI) outlets are installed in locations where moisture increases the risk of shock, such as in a bathroom or outside. When you try to plug an appliance into the outlet and the conditions are too damp, the GFCI shuts off. You'll need to push the reset button to make the outlet function. Test the GFCI monthly by pushing the test button. This action should cause the reset button to pop up. Push the reset button down so that the outlet functions.

Tasks

Replace fuses and reset breakers

If a light or plug will not work, you may have blown a fuse or tripped a breaker. You have to either replace the fuse or reset the breaker. Follow one of the methods discussed below.

Skill level rating: 1 - Simple maintenance

Materials: new fuses of the same type and rating

Tools: screwdriver, rubber-soled shoes, dry boards to stand on, flashlight

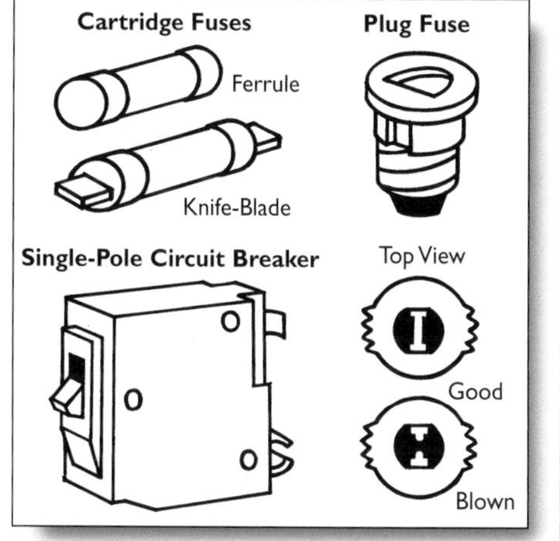

Replace plug fuses

1. Turn the main switch to OFF and put the shank of a screwdriver through the lock hole in the handle so the switch cannot accidentally be tripped to ON.
2. Open the fuse box cover. Look to see if the glass window on one of the fuses is blackened. If it is, a short-circuit has probably caused the fuse to blow. If the window is not blackened, look for a melted wire or loose spring end inside the glass window. This blown fuse "signature" means usually that the circuit is overloaded or has shorted out.
3. Check the circuit sketch or label inside the fuse box cover to see which outlets the blown fuse controls. If the fuse window is not blackened, move one or more high amperage appliances to an outlet on another circuit. Too many appliances drawing a lot of energy, such as toasters, air conditioners and refrigerators, probably caused the overload.
4. If the fuse is blackened, chances are that one of the devices operating off this circuit or the circuit itself has a short. In this case, switch all wall switches on the circuit to OFF and remove all electrical plugs from outlets on the circuit. Inspect the cords and plugs on all appliances that were running off this circuit for damage such as frays, bare wires or blackened plugs. Appliances in this condition must not be used again until they are repaired. If the appliances appear fine, there may be a short in the circuit.
5. Remove the blown fuse by turning it counter-clockwise. The number on the end of each fuse indicates its amperage rating. **CAUTION:** When replacing blown fuses, you must use a new fuse with the proper amperage for the circuit. Using a fuse with a larger number may lead to a fire.
6. Install a new fuse in the socket and turn it clockwise to tighten it.
7. Close the fuse box cover and turn the main switch to ON. If the fuse blows again immediately, leave it in the fuse box and call an electrician. Otherwise, plug in each visibly undamaged appliance one by one in a different circuit and turn it on briefly. If the new fuse blows when you turn on an appliance, remove that appliance and have it repaired.
8. Turn on each wall switch in similar fashion. If the fuse blows as you turn on a particular switch, the light or other device the switch controls is probably at fault. Tape the switch in the OFF position so it cannot accidentally be turned on before replacing the fuse again. Call an electrician to find and correct the problem.

Reset circuit breakers

1. Circuit breakers trip to one side or pop out to break the circuit. Check the labels in the circuit box to see which outlets the tripped breaker controls. It is not necessary to turn the main switch off.
2. Move one or more high-amperage appliances to another circuit, if you suspect the circuit is overloaded. Flip the circuit breaker to the other side.
3. If the circuit breaker trips again, switch all the wall switches on the circuit to OFF and remove all electrical plugs from outlets on the circuit. Reset the breaker and re-insert the plugs back into the outlets.
4. Now turn on each appliance briefly, one at a time, and check the breaker each time you add an appliance. Then turn the wall switches to ON, one at a time, checking the breaker. If it trips, then the appliance you have just plugged in or the device controlled by the switch you have just turned on is defective and must be repaired or replaced. If a switch is the problem, tape it in the OFF position so it cannot accidentally be turned on. Reset the breaker and call an electrician to find and correct the problem.

Replace cartridge fuses

Skill level rating: 4 - Qualified tradesperson/contractor

Cartridge fuses are used in the main switch box to control all electricity coming into your home. They may have round or knife-shaped end caps.

If cartridge fuses are blowing frequently in your home, you will either have to use fewer electrical devices or call an electrician to upgrade or correct your service. Replacing cartridge fuses can be very dangerous and is not a job for a homeowner.

Replace a wall outlet

Replace an outlet when it will no longer hold a plug securely, if it's not working properly or if you want to install a childproof outlet.

Skill level rating: 3 - Skilled homeowner

Materials: new outlet, new faceplate if required, masking tape

Tools: screwdriver, needle-nose pliers, flashlight

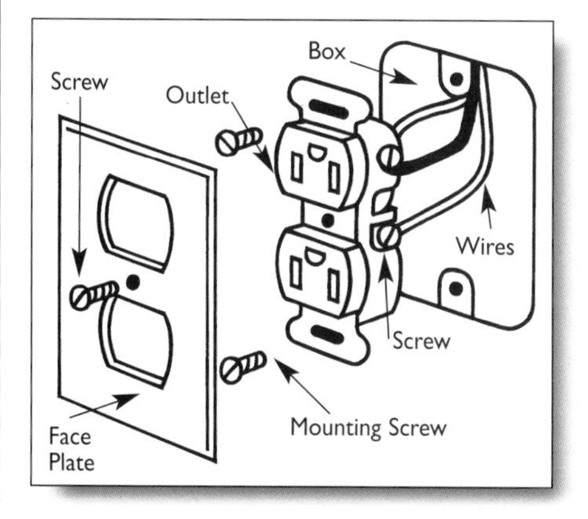

1. Turn off the circuit breaker or remove the fuse that controls the power to the outlet you are repairing before proceeding. Better still, turn the main switch to OFF and lock it OFF until your work is through.
2. Remove the faceplate screws and faceplate.
3. Remove the mounting screws.
4. Gently pull the outlet from the box. Attach a piece of masking tape to each wire and label it to help you to remember the terminal to which it connects.
5. Loosen the terminal screws and remove the wires.
   Note: Some outlets have terminal inserts in place of screws. These inserts have a release bar that you press with a screwdriver to release the wires.
6. Remove the old outlet. Reattach each wire around the proper terminal screw on the new outlet. The black wire attaches to the brass terminal. The white attaches to the silver. The end of the wire must face in the same direction that screws will turn when tightened. Tighten screws.
   Note: If the outlet has terminal inserts, just push the wires firmly into the inserts instead. Remove the labels.
7. Gently push the outlet back in the box. Install it with the mounting screws provided, complete with the isolating gasket. Modern outlets have ground or bare wires between the box and the outlet. These wires have to be attached to the proper terminal (green). Older systems may not have ground wires.
8. Replace the faceplate or install a new one.
9. Turn the power back on and plug in an appliance or lamp to test the new outlet.

Replace a wall switch

If a wall switch fails to work, it should be replaced. The switch faceplate may also need replacing.

Skill level rating: 3 - Skilled homeowner

Materials: new switch, new faceplate if required, masking tape

Tools: screwdriver, needle-nose pliers, flashlight.

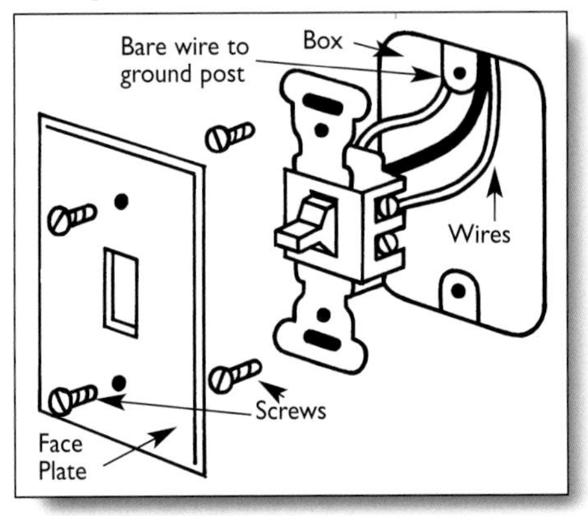

1. Turn off the circuit breaker or remove the fuse that controls the power to the outlet you are repairing before proceeding. Better still turn the main switch to OFF and lock it OFF until your work is through.
2. Remove the faceplate screws and faceplate.
3. Remove the two switch mounting screws.
4. Gently pull the switch from the box. Attach a piece of masking tape to each wire and label it to help you remember the terminal the wire connects to.
5. Loosen the terminal screws and remove the wires. Remove the switch.
6. Reattach each wire around the proper terminal screw on the new switch. The end of each wire should face in the same direction that screws turn when tightened. Tighten the screws firmly and remove the labels.
7. Gently push the switch into the box and secure it with the screws provided. Replace the old faceplate or use a new one.
8. Turn the power back on and trip the new switch to ON and OFF a few times to check its operation.

Replace a cord plug

Cord plugs need to be repaired when their wires come loose or are damaged. Don't use an appliance with a damaged plug.

Skill level rating: 3 - Skilled homeowner

Materials: new plug of the same type

Tools: screwdriver, knife, scissors or pliers

There are two main types of household cord plugs—flat and round. Flat-cord plugs have two prongs, are for light duty equipment and are the most common. Round-cord plugs have two and often three prongs and are for heavier equipment that requires grounding.

Tip: Always replace a plug with a plug that is similar in size and type.

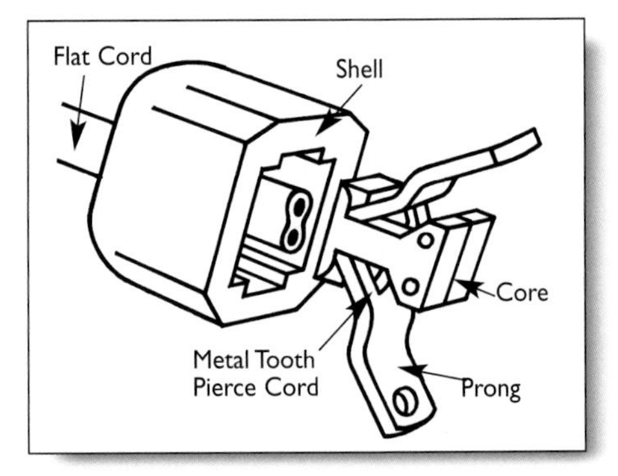

Replace a flat-cord plug—clamp-style

1. Cut the old plug from the cord.
2. Pull plug core from new plug. Spread prongs.
3. Insert end of cord through plug hole and into the core. If the plug is polarized, the neutral side will have to connect with the wide prong. The neutral side is ridged.
4. Squeeze prongs together, which will pierce the cord and establish a contact.
5. Slide core securely back into the plug case.

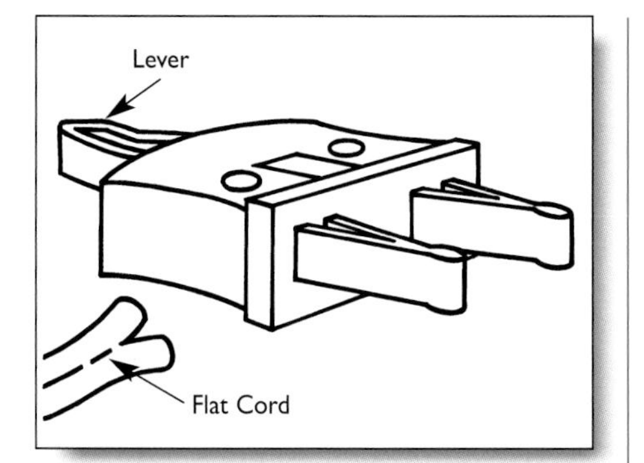

Replace a flat-cord plug—snap-style

1. Cut the old plug from the cord. Spread the wires apart slightly.
2. Lift the lever on the plug.
3. Insert the wire into the plug casing.
4. Close the lever to establish the contact.

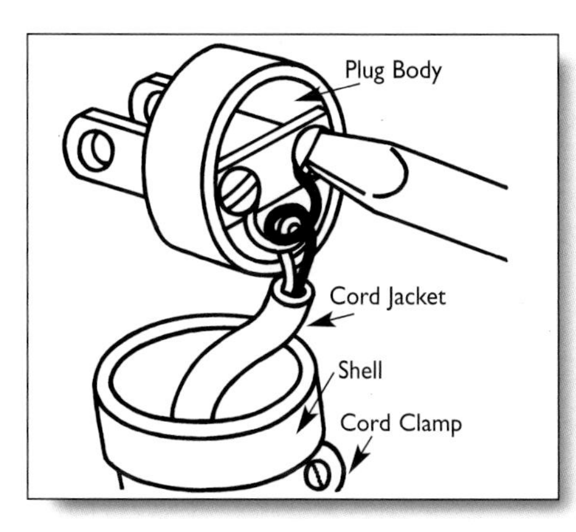

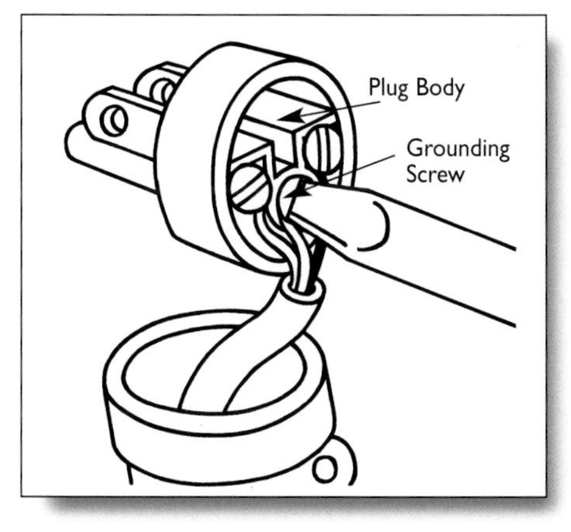

Replace a round-cord plug

1. Cut the old plug from the cord.
2. Pull the new plug core out of the plug case.
3. Push the cord through the centre hole in the new plug case. Carefully remove 30 mm (about 1 1/4 in. of the outside (the first layer) insulation from the end of the cable.
4. Strip the insulation underneath (the second layer) back 20 mm (about 3/4 in.) from the end of each wire to expose the bare wire underneath. Twist each bare wire end between your thumb and forefinger to hold the strands together.

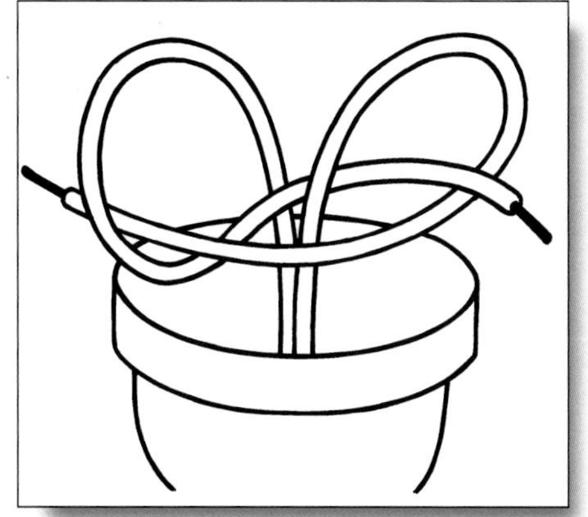

5. Tie an Underwriter's knot in the wires so they cannot be pulled loose. Pull the cord from the back of the plug until it stops. If the plug has three prongs, the bare or green wire (ground) must be attached to the green screw.
6. Place each wire around the correct terminal screw (white to silver screw; black to brass screw) so its end faces in the same direction that the screw will tighten when it is turned.
7. Tighten the screws and reinstall the plug core into the case.

Replace a simple light fixture

Skill level rating: 3 - Skilled homeowner

Materials: new fixture, marrettes

Tools: screwdriver, utility knife, wire strippers, needle-nose pliers, flashlight

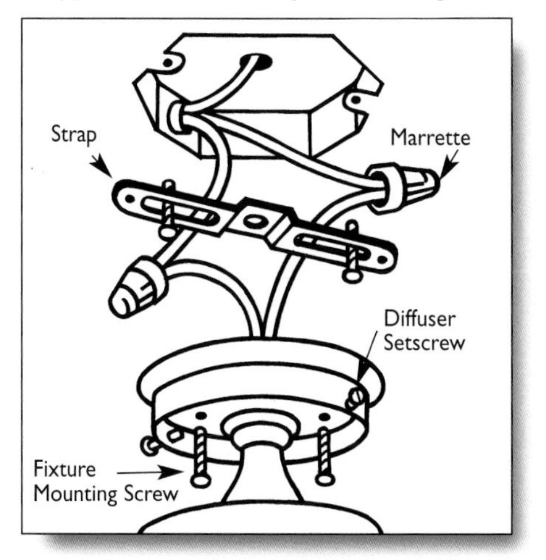

1. Turn off the circuit breaker or remove the fuse that controls the power to the fixture you are repairing before proceeding. Better still, turn the main switch to OFF and lock it OFF until your work is through.
2. Loosen the screw holding the fixture globe and remove the globe. Remove the light bulb. Free the fixture by either removing the mounting screws or turning and removing the fixture (bayonet fitting).
3. Twist off the marrettes, holding the wires together.
4. Untwist wires and completely remove the fixture.
5. Install the replacement fixture. The new fixture may have special installation instructions that should be followed. If not, pull wires through the new fixture and re-connect the wires using the marrettes.
6. Mount the fixture using the mounting screws or special mounting fitting.
7. Install the light bulb and globe, then tighten the screws to hold it in place.
8. Turn the power back on.

HEATING AND COOLING EQUIPMENT

Maintaining heating and cooling equipment saves you money on energy costs, improves comfort, helps preserve the energy efficiency and durability of your house and protects your family's safety.

The most common heating and cooling equipment problems are:

- poor air flow through central forced air heating and cooling system ducts
- noisy or poor circulation in hot water heating systems
- poor thermostat operation
- burning dust odours from electric baseboard heaters
- combustion spillage
- chimney obstructions
- fires
- inoperable smoke alarms or CO (carbon monoxide) detectors
- clogged furnace filters
- dirty or moldy central furnace-mounted humidifiers
- dirty or moldy air conditioning coils or drip pans
- unsafe wood stove or fireplace conditions

Maintenance includes:

- Forced air systems—removing obstructions from registers and accessible portions of ducts and ensuring that all ducts are attached.
- Hot water systems—checking for any leaks.
- Electric (baseboard systems)—cleaning and checking attachment.
- Furnace maintenance and safety—filter replacement, cleaning humidifiers and air conditioning coils.
- Fire safety—checking periodically for combustion safety, dealing with fires, checking and maintaining smoke alarms and CO detectors.
- Wood stoves and fireplaces—regular chimney cleaning, ash removal, keeping surrounding areas away from combustible materials

You can do many of the periodic heating, and cooling equipment maintenance tasks. However, it is essential to have any combustion equipment (using any form of gas, oil or wood) serviced once a year by a qualified professional.

Prevention tips

- Start with an annual maintenance routine performed by a professional serviceperson for all your combustion appliances. The service person should check for heat exchanger leakage, evidence of start-up spillage and chimney condition. Annual maintenance should include a tune-up because a properly tuned combustion appliance rarely produces carbon monoxide.

- The CMHC garbage bag airflow test is one simple test that the homeowner can do to estimate airflow out of registers or exhaust fans.

CMHC garbage bag airflow test

Skill level rating: 1 - Simple maintenance

Materials: 1 wire coat hanger, 1 large garbage bag, 1 large 1.2-m (4-ft.) long leaf collection bag, tape

Create a garbage bag tester

1. Using two hands, grasp the coat hanger by the hook and by the base.
2. Pull apart enough to create a rectangular wire hoop.
3. Tape the wire to the mouth of the garbage bag or leaf collection bag to keep it open.

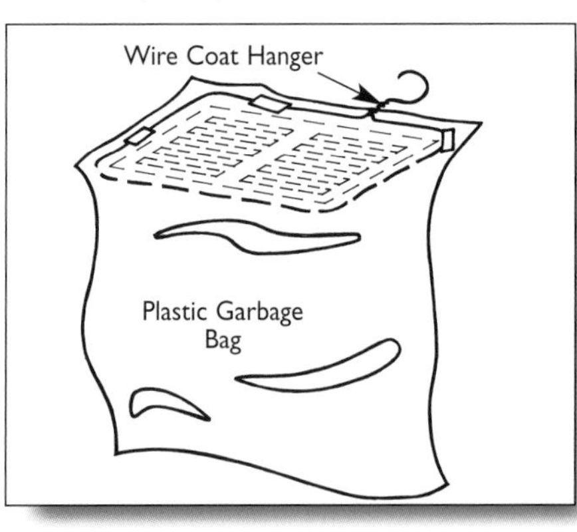

Test a bathroom fan with the large household garbage bag:

1. Turn the fan on.
2. Swing the bag to inflate it and hold the opening tightly to the fan opening. The exhaust flow will deflate the bag.
3. Time how long it takes to deflate.

Time to deflate	Airflow
3 seconds	good
5–6 seconds	mediocre
10–12 seconds	poor

Test a heat recovery ventilator with a leaf collection bag

1. With the heat recovery ventilator (HRV) operating, go outside to where your HRV ducts exit the house.
2. Crush the bag flat and hold the opening tightly over the exhaust hood. The air flowing out of the hood will inflate the bag.
3. Time the inflation. If the bag inflates in less than eight seconds, turn the HRV to a lower speed, then repeat the test. Take note of the time.
4. Now swing the bag to inflate it and hold the opening against the wall around the HRV supply hood. The air going into the HRV will deflate the bag.
5. Time the deflation. If the HRV is balanced, the inflation and deflation times should be roughly equal. If, for example, one time is twice as high as the other, the HRV is unbalanced. If so, check each of the maintenance steps below and try again. If it still seems unbalanced, call a service person to test and adjust your HRV.

Repair tips

- If your oil furnace starts but stops shortly after it starts, check to make sure there is fuel in the tank before calling the serviceperson.

Special considerations

Healthy Housing™

- Furnaces typically last 15 to 20 years, whereas boilers may last up to 40 years if maintained properly. New heating and cooling equipment is much more energy efficient than the older equipment that you may have in your home.

- Upgrading insulation and draftproofing your home may reduce heating and cooling needs.

Safety

- Clogged filters and flues are a fire hazard.

- Incomplete combustion (burning) of fuels, backdrafting (reversed air flow through flues or combustion appliances) of combustion gases or cracked heat exchangers may allow combustion gases into the home's air. **CAUTION:** This is a serious health and safety risk to the occupants and demands immediate attention.

- Ensure there are no drapes or objects that could start a fire if they come into contact with an electric baseboard heater.

Tasks

Clear obstructed ducts and registers

Forced air heating systems have ducts that carry the air to and from the rooms in your house. People are often concerned about how to clean out dust and debris from inside the ducts. Duct cleaning research has indicated that there is little or no difference in the concentrations of house airborne particles or in duct airflows following professional duct cleaning. Regular, professional duct cleaning is unnecessary. However, in some instances professional help makes sense:

- If you have a problem with moisture or water in your ducts, you may have mold in your ducts as well. Cleaning and prevention of further moisture or water entry is essential. Replacing the affected duct section might be necessary.
- If you are moving into a newly constructed or newly renovated house.
- If you are having a new furnace installed that has a more powerful fan than your current furnace fan.

It is important to keep the ducts and registers clear from obstructions so that the air flows smoothly to and from the rooms in your home.

Skill level rating: 2 - Handy homeowner

Materials: possibly sheet metal screws to secure duct sections, aluminum foil duct tape or water-based duct sealer

Tools: screwdriver, household vacuum cleaner, flashlight, small mirror, large garbage bag taped to a coat hanger hoop, possibly portable drill and small brush.

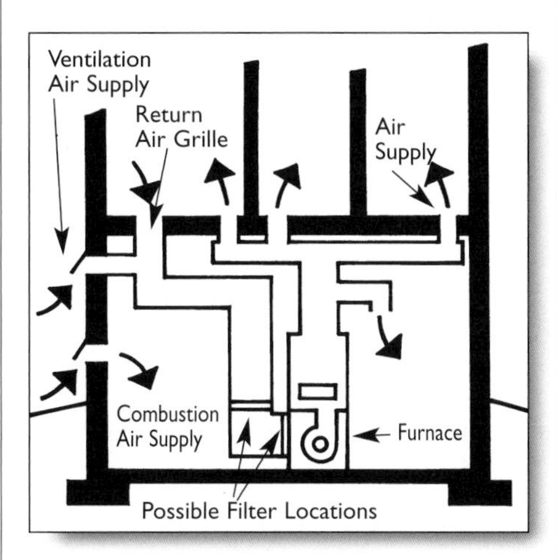

1. Locate all air supply and return registers. Move furniture or rugs that may be obstructing airflow from registers.
2. Remove the register at each outlet.
3. Discard any filters at the registers. They don't reduce breathable dust and they can obstruct airflow.
4. Inspect the portion of the duct that you can easily see from the register. Use a flashlight and mirror to get a better look.
5. Remove any large objects by hand.
6. Use a vacuum cleaner to remove any small debris.
7. Check that the duct sections are securely connected to the floor framing under the register. Wherever possible, check that there are no loose joints between the duct sections. If necessary and where accessible, screw sections together.
8. Seal all accessible duct joints with tape or duct sealant.
9. Replace registers. Ensure that any moveable register louvres are fully open.
10. Use the CMHC garbage bag airflow test to assess airflows to individual rooms with room doors both open and closed. If airflows are uniformly low, in-duct dampers may need adjustment or a professional may need to assess the system capacity and balance airflows to all rooms. If airflows are substantially reduced only when doors are closed, undercutting doors to permit air movement may be required.

Fix noise or poor circulation in hot water heating systems

Air is present in hot water heating systems in the form of dissolved air, entrained (trapped) air in the form of tiny bubbles or actual larger bubbles. Sometimes air in the system can cause gurgling noises or interfere with water circulation and heating capability.

Baseboard and cabinet convector systems have automatic vent valves to resolve this problem. Convectors are "today's radiator." They can be baseboards or cabinet style and have copper or steel tubing, surrounded by metal fins. This system is much more efficient than the old, cast iron radiator.

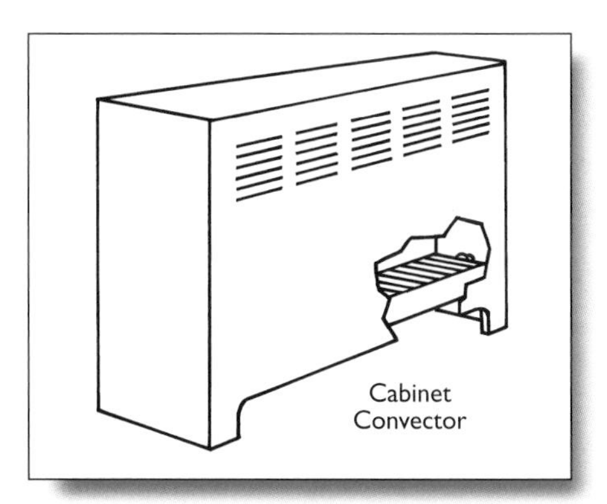

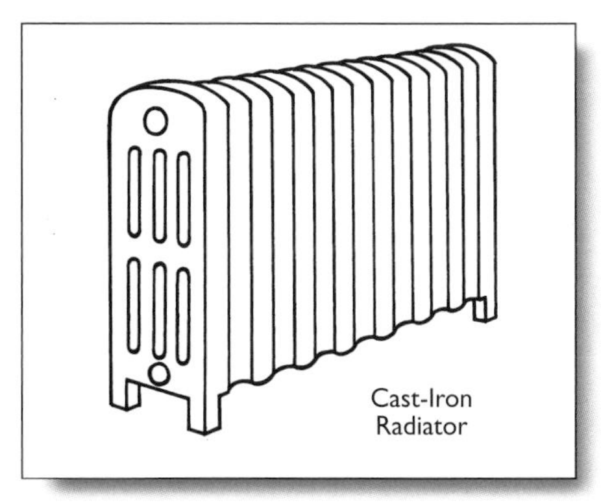

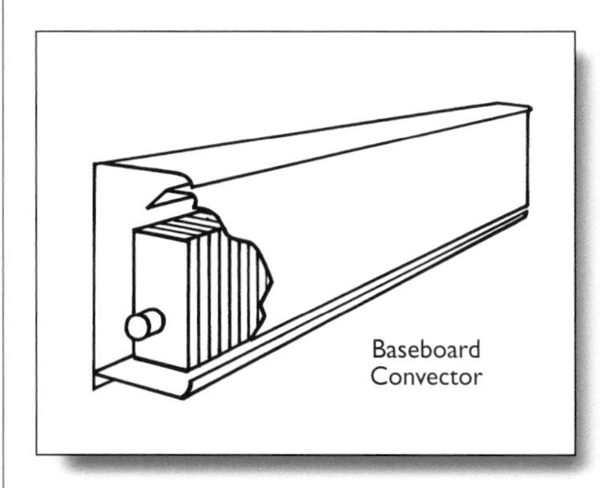

Keep automatic vents clean. If they start to leak water, that is a signal for replacement. However, some convectors, as well as many of the older cast iron radiator systems, may need manual venting, particularly when systems have been turned on for the first time in a long while.

Skill level rating: 2 - Handy homeowner

Materials: none required

Tools: a slot screwdriver and a cup or other small container

If more than one radiator or convector seems cool, start with one on the lowest level of your house.

1. Find the vent valve. On a radiator, it is near the top, opposite the inlet valve. On a convector, remove the cover to find the valve.
2. Turn the screw on the top of the valve counter-clockwise to open.
3. Leave it open until water spurts out; be ready to catch the hot(!) water in the cup.
4. Close the valve.

Replace a thermostat

A thermostat switches your heating and cooling system on and off, based upon the temperatures you've set. It should be installed on an inside wall in a draft-free spot. If your thermostat is old and not working properly, you may need to replace it. Or, you may want to upgrade to a more modern one that has more features.

Skill level rating: 3 - Skilled homeowner

Materials: replacement thermostat that is compatible with your heating system

Tools: screwdriver, level, pencil, wire strippers,

1. Turn off the power to your heating system and thermostat.

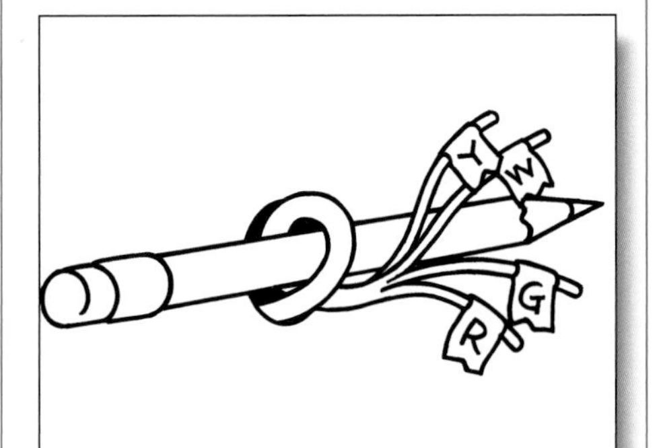

2. Remove the cover of the old thermostat. Unscrew the unit. Disconnect the wires from the unit.

3. Slip the wires through the new unit. Position the unit straight on the wall and mark the mounting holes. You may want to check the positioning with a level so that it will be straight when you attach it to the wall. Screw the unit into the wall using your marks as a guide.

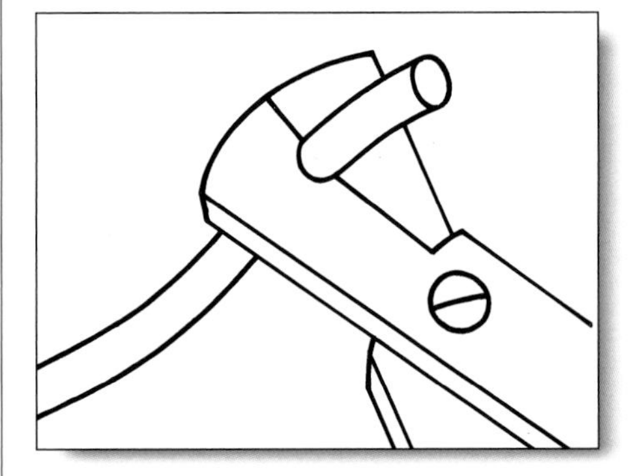

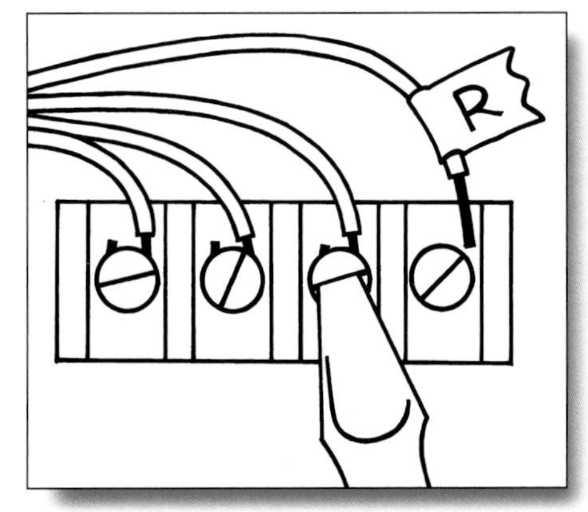

4. Strip 10 mm (3/8 in.) of insulation from the end of the wires. Connect the wires to the unit, following the manufacturer's instructions. Push any excess wire through the hole and into the wall.

5. Attach the front cover.
6. Turn the power back on. Program the thermostat to your desired settings.

Attach loose electric baseboard heaters

Sometimes electric baseboard heaters can inadvertently become detached from the wall.

Skill level rating: 2 - Handy homeowner

Materials: 1 1/2" flat head wood screws

Tools: Screwdriver

1. Make sure that the thermostat is turned right down.
   CAUTION: It is a good idea to turn off the power to the heater at the electrical panel.
2. Remove the front heater cover to expose the metal fin convectors. Do not tamper with the covers or electrical connections at either end of the heater.
3. Most baseboard heaters have a series of attachment holes through the back of the housing. Tighten the existing screws through these attachment holes, if possible.
4. If necessary, insert new screws through other attachment holes and into the wall studs. Attach the heater at least an inch off the floor. On most walls, you are likely to find studs about 400 mm (16 in.) away from the original screw placements.
5. Check to make sure that the heater is firmly attached.
6. Replace the cover by hooking it on the top hooks and pushing gently to engage the bottom hooks.
7. Turn on the power to the heater at the panel. Turn the thermostat to the desired setting.

Clean electric baseboard heaters

When electric baseboard heaters are first turned on after a long while, there may be a smell of burning dust.

Skill level rating: 1 - Simple maintenance

Materials: none

Tools: typical home vacuum cleaner with small nozzle attachment

1. Before turning on the electric heat every year, clean every heater. Make sure the thermostat is turned right down.
2. Remove the front heater cover to expose the metal fin convectors. Do not tamper with the covers or electrical connections at either end of the heater.
3. Gently vacuum the metal fins.
4. Replace the cover by hooking it on the top hooks and pushing gently to engage the bottom hooks.

Inspect furnace for signs of inefficient combustion

Make sure that you have a homeowner's manual for every combustion appliance in your home. Read and follow the manufacturer's safety and maintenance instructions. Most manuals will include a homeowner's recommended maintenance schedule.

Many manufacturers recommend a monthly visual inspection of the combustion appliance and venting, looking for soot accumulation on or near the flue pipe or dampers, unusual moisture, corrosion or paint discoloration. Any of these signs may indicate that the by-products of combustion (which may include deadly carbon monoxide (CO)) are spilling back into the home instead of going out the chimney or vent. If you have any concerns, call your serviceperson.

Manufacturers often suggest monthly inspection of the main burner flame in a natural gas furnace. This could require opening an inspection port. Check the flame making sure it is blue with orange streaks. A bright yellow flame indicates incomplete combustion. Call your service person if the flame is yellow.

Skill level rating: 1 - Simple maintenance

Materials: none

Tools: manufacturer's maintenance checklist

Identify and prevent combustion spillage

Any fuel-burning device produces combustion gases that can include toxic elements such as carbon monoxide. Most natural gas, oil and propane equipment produces little carbon monoxide if properly installed and maintained. Burning wood, kerosene, coal or charcoal produces carbon monoxide. Even at low levels of exposure, carbon monoxide can cause serious health problems. Normally, combustion gases are vented to the outdoors through a chimney or vent pipe. However, due to chimney problems, equipment problems or air pressure problems, combustion gases may escape into your home. Combustion spillage is the term used to describe the unwanted flow of combustion gases into your home.

- Have a qualified service person inspect and clean fuel burning appliances and venting annually. The service person should check for heat exchanger leakage and evidence of start-up spillage and should check the chimney. Install detection devices to alert you if spillage is occurring. There should be a smoke alarm installed on the ceiling near a fireplace or on the ceiling above the damper of an oil furnace. There should also be a CO detector installed near any combustion appliance that is vented using a chimney.
- Avoid operating several powerful exhaust devices simultaneously. These could cause combustion appliances to backdraft.
- If you install a new range-top grill with a powerful exhaust fan, get expert advice on how to supply sufficient air to the house while it is operating.
- If your furnace or water heater is enclosed in a small separate room, allow air to move freely between the furnace room and the rest of the house (through louvred doors for instance).
- If you have a forced air heating

system, be sure you are not drawing return air from the immediate vicinity of your combustion appliances. Make sure the blower door on your furnace is in place.

- If you have a gas range, be sure to use a range hood that exhausts to the outside while the range is operating.
- Watch for warning signs of combustion spillage:
  - repeated headaches, skin and throat irritations, and other low grade illnesses
  - combustion odours anywhere in the house
  - hot and muggy air around the furnace
  - soot stains around any combustion appliance, or unusual rumbling sounds when it is operating.

What to do if you smell gas

Safety

Fire or explosion may result from improperly installed or maintained natural gas appliances. It is also important not to store gasoline or other liquids with flammable vapours near any combustion appliance.

If you smell gas:

- Do not try to light any appliance.
- Do not touch any electrical switch; do not use any telephone in your building.
- Get everyone out of the house.
- Immediately call your gas supplier from a neighbour's telephone. Follow the gas supplier's instructions.
- If you cannot reach your gas supplier, call the fire department.

Install and maintain carbon monoxide detectors

The best way to protect against health risks from carbon monoxide (CO) is to eliminate sources of CO in the home. Combustion appliances are not the only possible sources of CO in your home. Research on attached garages has indicated that house/garage connections may leak roughly as much air as the rest of the house envelope. Garage/house air exchange is significant and can be an entry point for pollutants, including CO, from the garage and vehicles.

Carbon monoxide safety in garages

- Never start a vehicle in a closed garage; open the garage doors first. Pull the car out immediately onto the driveway, then close the garage door to prevent the exhaust fumes from being drawn into the house.
- Do not use a remote automobile starter when the car is in the garage. Even if the garage doors are open, CO may seep into the house.
- Do not operate propane, natural gas or charcoal barbecue grills indoors or in an attached garage.
- Avoid using a kerosene space heater indoors or in a garage.

Skill level rating: 2 - Handy homeowner

Materials: CO detector complete with screws and mounting instructions

Tools: screwdriver

Install a CO detector where you will hear it while sleeping. A detector can be placed at any height, in almost any location, as long its alarm can be heard. To avoid damage to the unit and to reduce false alarms, do not install a CO detector in unheated areas; in high humidity areas; where it will be exposed to chemical solvents; near vents, flues or chimneys; or within two metres (six feet) of heating and cooking appliances.

Most CO detectors are designed to alarm when CO concentrations reach a high level in a short time. However, long-term, low-level exposures are also of concern, especially for the unborn, young children, the elderly and those with a history of heart or respiratory problems. Choose a CO detector that is listed with ULC (Underwriters' Laboratory of Canada) to the CGA (Canadian Gas Association) or CSA (Canadian Standards Association) standard. If you want to monitor long-term, low-level exposure and short-term, high-level exposure, choose a unit with an adjustable display and a memory. Battery-operated units can be conveniently located but require the user's diligence in replacing worn-out batteries. Plug-in units should not be connected to an electrical outlet controlled by a switch. Replace any detector at least every five years.

Deal with fires

Naturally, preventing fires is best. Reduce stored items in the house or garage, particularly flammable liquids like gasoline and solvents. Eliminate clutter so fire will have fewer places to start. Check smoke alarms and CO detectors regularly to ensure that they are in working condition and that their batteries have not worn out. Make sure the whole family is aware of emergency exit plans for fires in various locations in the home. Make sure that any renovations to create bedrooms, particularly in basements, follow the local building code regarding means of escape from fires. Keep general-purpose ABC fire extinguishers handy and serviced.

If a serious fire breaks out, make sure that the fire doesn't trap anyone from reaching an exit. Leave quickly, closing doors as you go to slow the fire's spread. Feel interior doors for heat before opening them. If necessary, find another way out. Smoke rises, so stay low.

Put out a small fire

Skill level rating: 1 - Simple maintenance

Materials: none

Tools: ABC general-purpose extinguisher, water, baking soda or salt

1. Remove the locking pin on the fire extinguisher.
2. Aim at the base of the fire.
3. Pull the trigger on the fire extinguisher.
4. Sweep back and forth until the fire is out.

5. If you don't have an extinguisher, use water to put out fires on most household items like cloth, paper or wood.
6. Either smother or sprinkle baking soda or salt on grease fires.
7. Cut off the power to an electrical fire and use an extinguisher that contains a C rating.

Chimney fires

Skill level rating: 1 - Simple maintenance

Materials: none

Tools: none

1. Call the fire department.
2. Close off the air supply to the combustion appliance.
3. If there is a barometric damper on the pipe connecting a furnace and chimney, make sure that this damper stays closed.
4. Do not try to pour water into the chimney. The resulting steam can be very dangerous and the shock of the cold water hitting the hot fire can destroy the chimney.
5. Have the chimney inspected thoroughly after having a chimney fire.

Check for damaged and unsecured chimneys

Damaged chimneys may contribute to combustion spillage. Obstructions such as birds' nests, broken bricks or flue liners, or ice can block air flow through a chimney. Damaged or unsecured chimneys, particularly those made of brick or block, may collapse, creating a severe hazard to a passer-by.

Skill level rating: 2 - Handy homeowner

Materials: none

Tools: checklist, mirror, flashlight or emergency light and rope, binoculars, ladder, personal safety equipment

1. As part of the seasonal inspection, check the condition of the chimney. If you have any doubts about the condition of your chimney, arrange for a professional inspection (which should always be done as part of your annual combustion appliance service). Inspect the chimney from the exterior and from the interior, if possible.
2. For the exterior check, look for leaning, cracks, loose mortar or dislodged brick in masonry chimneys, and the presence and condition of chimney caps. Binoculars are often helpful.
3. A bright, sunny day that will make it easier to see as you do the interior check. Make sure the appliance is not operating until the inspection is complete. Remove the bottom cap from factory-built chimneys or the clean-out door from masonry chimneys. Using a light and a mirror, try to assess the interior chimney condition. Look for any obstructions or creosote build-up. Significant debris in the clean out is a sign of a deteriorating masonry chimney.
4. If the interior check is not possible, you may need to access the top of the chimney and lower a light down on a rope. Be careful! Use appropriate safety precautions and gear such as a harness if you are working on the roof.

Clean chimneys

Unburned components of the smoke passing up a chimney can condense on the inside of the lining and can cause fires. If your chimney inspection reveals 6 mm (1/4 in.) or more of dusty or flaky black residue, your chimney should be cleaned. If the residue is very hard or has a glazed look, professional cleaning might be needed.

Chimney cleaning is a messy job, so be prepared. Masonry chimneys in particular may need to be cleaned from the top down.

CAUTION: Any work involving ladders and roofs is dangerous. Be careful!

Skill level rating: 2 - Handy homeowner or
Skill level rating: 4 - Qualified tradesperson/contractor (WETT certified professional)

Materials: none

Tools: a chimney brush to match the chimney type and size; chimney brush extension rods; goggles; dust mask; old clothes or throwaway coveralls; gloves; plastic sheeting to protect area at the base of the chimney; perhaps a ladder; safety equipment; mirror and light for inspection; small brush, scoop and pail; shop vacuum.

Whether working up or down, any stovepipes that connect the combustion appliance to the chimney should be carefully removed and cleaned (preferably outside). Ensure that the stovepipes are in good condition before they are securely reinstalled after the chimney cleaning.

To work up:

1. The preferred approach is to work from the bottom up, if possible. Seal any openings into the house such as fireplaces, base tees or chimney cleanouts.
2. To work up through a masonry fireplace: open the damper; insert the chimney brush and first section of extension rod; pass the end of the rod through a hole in a plastic sheet; and tape the plastic sheet to the face of the fireplace to contain the soot. For a metal chimney with a base cap: remove the base cap and insert the brush and first extension rod. Use the plastic sheeting to shield surrounding areas.
3. Use the brush to scrub the flue, adding extension rods as needed.
4. Remove the plastic sheet and unseal the openings.
5. Carefully gather the soot with the brush, scoop and pail.
6. Vacuum the remaining soot.
7. Inspect the chimney. It should be clean and undamaged.

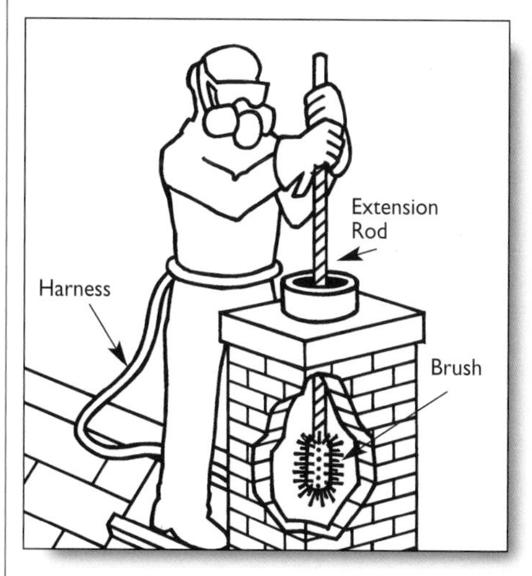

To work down:

1. Seal any openings into the house such as fireplaces, base tees or chimney cleanouts.
2. Using proper safety precautions and safety gear, carefully access the top of the chimney.

3. Remove the chimney cap if required.
4. Use the brush to scrub the flue from the top down, adding extension rods as needed.
5. When you are finished, remove the interior covers; gather the soot; vacuum as required.

Clean or replace furnace filters

There are many different types of furnace filters. Some are permanent and can be cleaned by removing, vacuuming or washing with hot water and detergent and replacing them. Other filters are periodically discarded and replaced.

Some filters are electronic or are in the same compartment as operating blowers.

CAUTION: Be careful to follow the manufacturer's instructions before opening any filter compartment or case, to avoid electrical shock or other hazards.

Healthy Housing™

Traditionally, furnace filters were designed to protect the furnace and fans. Some filters are now being installed to reduce exposure to particles that can affect your health. Research indicates that people's exposure to airborne particles appears to be directly linked to their activities at home. The furnace filter appears to have only a moderate effect on the exposure of an individual to respirable particles in the home. However, it can do a good job of keeping the air clean in the ducts.

To upgrade from the typical 25 mm (1 in.) glass fibre furnace filter, your choices include:

- 25 mm (1 in.) pleated filter
- 25 mm (1 in.) premium pleated filter (with electrostatic charge)
- charged media electronic filter
- 100 mm (4 in.) pleated media filter
- one of the high-efficiency bypass filters such as HEPA (high efficiency particulate air)
- an electronic plate and wire (ESP—electrostatic precipitator) type

Ultimately, your decision to change to a different type filter may depend upon whether the new filter will fit into the existing slot in your duct system or whether you wish to have the return air trunk professionally adapted to accommodate the type of filter that you prefer.

Research indicates that the 25 mm (1 in.) premium pleated filter provides good filtration and cost effectiveness for the amount of clean air delivered in comparison to other filters that will fit into the same slot. Overall, ESP filters seem to be most effective in terms of clean air delivered and cost per amount of filtered air. However, they produce small amounts of ozone and other respiratory irritants.

Cleaning or replacing 25 mm (1 in.) furnace filter located in return duct slot

Skill level rating: 1 - Simple maintenance

Materials: unscented dish soap and warm water for cleanable filter; replacement filter for throw-aways

Tools: none

1. Turn the thermostat right down so the furnace won't start.
2. Find the furnace filter. It is often in a slot in the return air duct just before it enters the side of the furnace.
3. Grasp the edge of the filter and slide it out.
4. If it is a metal or heavy polypropylene mesh filter, it will usually have some indication that it is washable. If so, wash it in warm soapy water, rinse and allow it to dry.
5. If it is a throw-away type, discard it and replace it with the same size filter from the available types.
6. Set the thermostat back to the desired temperature.

Cleaning or replacing a washable filter located in the blower cabinet

1. Turn off the power to the furnace.
2. Open the blower cabinet by following the manufacturer's instructions. Many furnaces have a panel on the front at the bottom that slides up and out or is fixed with a couple of screws.
3. Locate the filter. Remove, clean and reinstall. If the filter can't be cleaned, replace it with a new filter.
4. Close the blower cabinet.
5. Restore power to the furnace.

Humidity in houses

In newer homes that are more tightly sealed, daily living usually provides enough humidity to be comfortable. Too much humidity can often lead to problems such as water condensing on cold surfaces or molds growing inside the house. In older homes, homes with fewer occupants or homes in drier regions, sometimes the relative humidity level is too low for comfort. Before adding moisture to the air, it is a good idea to measure the relative humidity with a simple hygrometer that can be purchased at a hardware store. The sensation of dry air may be due to poor air quality and not low relative humidity.

Control humidity and eliminate mold to breathe easier. Do not humidify without measuring the relative humidity level first to determine if the house is too dry. Many houses do not need extra moisture. A relative humidity of 30 per cent in the winter should be sufficient to avoid breathing or mold problems. Relative humidity above 50 per cent next to cold surfaces can lead to mold growth. (Note: the relative humidity next to cold surfaces is higher than that in the middle of the room)

Clean central furnace-mounted humidifiers

Humidifiers can be useful for increasing moisture in the air. However, humidifiers need to be kept clean so they don't clog up with minerals from the water or have mold growing in them. Evaporative humidifiers are common. Typically, some of the air passing through the duct system is diverted through a humidifier mounted on the side of the furnace. The humidifier often consists of some type of wick or rotating sponge that picks up water from a pan or nozzle. Although this section deals with furnace-mounted humidifiers, you should also clean stand-alone room humidifiers frequently—possibly each time you fill them—but do so according to the manufacturer's instructions.

Skill level rating: 1 - Simple maintenance

Materials: replacement sponge or wick

Tools: screwdriver, small scrub brush, unscented household cleaner

1. Shut off the water supply to the humidifier. Usually there is a small shut-off on the water supply line to the unit.
2. Follow the manufacturer's instructions, if available.
3. Most furnace-mounted humidifiers have a cover attached with one or two hand-screws. Lift or remove the cover.
4. Slide out the water reservoir pan. Clean with detergent and water.
5. To replace the wicks or drum sponge, slide the old one out and slide the new one in.
6. Replace the pan.
7. Replace the cover.
8. Turn on the water supply.

Clean and service air conditioner

Have a professional service the air conditioner compressor, do a refrigerant check and clean the air conditioning coils and the drain.

Maintain wood stoves and fireplaces

Wood-burning appliances can heat a home. Fireplaces and wood stoves can be aesthetically pleasing. New designs can burn wood more efficiently. New standards and installer training programs allow you to get safe installations and dependable service.

New appliances should carry either a CSA, ULC or Warnock Hersey label certifying that the unit has been tested to established safety standards. The label will also advise you how far to keep combustibles away from the appliance. Installation and service professionals should carry certification from the Wood Energy Technical Training (WETT) Program.

Safety is a prime consideration with any combustion appliance. Hire a professional for an annual inspection and service program. Burning wet wood or reducing air supply to keep only smouldering fires can cause creosote to build up in the chimney, which could cause a chimney fire. Too much air supply and too much fuel in the firebox can overheat the appliance and chimney, creating an unsafe condition. Good burning techniques, dry wood, a clean wood-burning system and no combustibles near the appliance are the keys to safe, efficient operation.

Woodstove maintenance

Skill level rating: 1 - Simple maintenance

Materials: replacement firebricks

Tools: ash scoop, metal bucket, glass cleaner

1. Make sure that the fire is out and the unit is cool.
2. Regularly scoop ashes out into a metal bucket. Remove the bucket to a safe place outside, well away from the house or deck.
3. Check the condition of the firebricks inside the unit. They should be undamaged and in place. Replace any damaged or missing bricks.
4. Clean door glass with glass cleaner and a rag or paper towel. Never use glass cleaner or water on hot surfaces.

Firewood storage

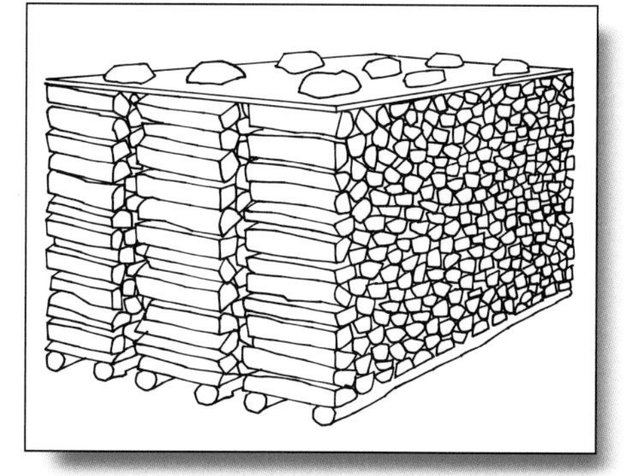

1. Stack firewood so that it doesn't rest directly on the ground and is covered by a roof or tarp. Leave the sides open for better air circulation.
2. Never store large amounts of firewood in the house. Drying wood can contribute to excess moisture and mold growth.

VENTILATION EQUIPMENT

We need to supply fresh air to our homes and remove stale air, odours and excess moisture. At some times of the year, opening windows is a low-cost way to reach these goals. But in Canada, that is not always practical. Having mechanical ventilation available can be effective and convenient.

There are several options. Kitchen and bathroom fans exhaust air from the places of highest odour and humidity, but they do not supply fresh air to the house. Centrally located supply and exhaust fans with branch ducts to various rooms are another choice.

The best alternative is usually a whole house heat recovery ventilation system. In this system, stale, humid air is exhausted directly from the kitchen and bathrooms through the heat recovery ventilator (HRV). An equal amount of fresh air is supplied through the heat recovery ventilator, where it recovers heat but does not mix with the stale air being exhausted. The fresh pre-heated air is then distributed throughout the building using either a dedicated duct system or the existing space heating duct system. A small amount of continuous ventilation can help to dilute pollutants and control humidity.

Too little humidity may be uncomfortable. This can sometimes occur during the heating season in older, drafty homes or in dry regions of the country. However, too much humidity, particularly in bathrooms, is a more common problem. This can lead to condensation on windows or other cold surfaces, mold growth and a favourable environment for dust mites. Mold and dust mites are common symptoms of poor indoor air quality and may cause respiratory and other health problems.

Ideally, relative humidity levels remain between 30 and 45 per cent. When it is below minus 10° C outside, surfaces will be colder and more prone to condensation, so the 30 per cent level (or lower, if necessary) is best in the winter. In the summer, it is sometimes difficult to keep relative humidity levels below 45 per cent. When the air outside is hot and humid, the best approach is to maintain some ventilation as needed for fresh air; keep basement windows closed to prevent the outside humid air from coming in and condensing on the cooler surfaces; and use a dehumidifier.

Insufficient mechanical ventilation is another cause of poor indoor air quality. Sufficient mechanical ventilation is a feature of housing that is healthy for the occupants and helpful for the durability of the building.

The most common mechanical ventilation problem is insufficient airflow caused by:

- dirty filters and screens
- inefficient duct systems

Maintenance includes:

Bathroom and kitchen fans

- cleaning fans
- cleaning range hood filters

Heat Recovery Ventilator

- cleaning the supply air intake screen, filter and core
- cleaning the condensate drain loop and discharge
- air sealing air and vapour barriers

Prevention tips

- Check the airflow from registers and exhausts using the CMHC garbage bag airflow test (see page 62). If you have kitchen and bathroom fans or an HRV and the air in your home still seems stale or damp, perhaps you have insufficient airflow. Some bathroom fans make a lot of noise but don't exhaust much air. HRVs should be balanced, with the fresh airflow matching the exhaust flow.
- Measure the relative humidity in your home and confirm whether the level is too high or too low. Use a mechanical or an electronic hygrometer. These are small, inexpensive and easy-to-use. Electronics stores or hardware stores commonly sell mechanical hygrometers for under $20 and electronic hygrometers for about $35 to $65.

Measuring humidity with a hygrometer

Skill level rating: 1 - Simple maintenance

Materials: none

Tools: mechanical or electronic hygrometer

1. Following the manufacturer's instructions for use. Place the hygrometer where the humidity symptoms are most obvious, usually in the room of greatest concern or where the occupants spend most of their time.
2. Try to place the hygrometer where it will be unaffected by direct heat, away from hot lights, radiators, heat registers or chimneys.
3. Leave the hygrometer long enough, according to the manufacturer's literature, to provide a stable reading.
4. Take readings in various rooms and levels of the house to get a sense of problem areas.
5. Use the information to decide whether to try to raise or lower humidity levels.

Keep your hygrometer accurate

Hygrometers need to be checked once a year for accuracy (calibrated) and sometimes adjusted so that they will make accurate readings. You cannot adjust an electronic hygrometer, but you can still calibrate it.

Skill level rating: 1 - Simple maintenance

Materials: 125 ml (about 1/2 C) table salt, 50 ml (about 1/4 C) tap water

Tools: mechanical or electronic hygrometer, coffee cup, large zip-lock plastic bag or a well-sealed pressure cooker.

1. Place the tap water and the table salt in the coffee cup and stir for about a minute.
2. Put the coffee cup and your hygrometer inside the plastic bag or pressure cooker and seal tightly. Note: Do not let your hygrometer come into contact with the salty water as it can damage your hygrometer.
3. Put the bag or pressure cooker in a draft-free place and out of direct sunlight. Select a spot where the room temperature is likely to remain even.
4. After 8 to 12 hours, note your hygrometer's RH reading. Your hygrometer should read about 75 per cent, the standard. If it does, you do not need to adjust it. If it does not read close to 75 per cent, record the difference between your hygrometer reading and 75 per cent.
5. If your hygrometer is adjustable, locate the adjustment screw or knob and immediately adjust the reading to 75 per cent. If your hygrometer is not adjustable, each time you take a reading you will need to subtract or add the difference that you noted.

Repair tips

- Calibrate your hygrometer once a year to ensure that the reading is accurate.

Special considerations

Healthy Housing™

- Household air can contain pollutants such as dust; combustion gases; cigarette smoke; chemicals off-gassed from building materials, furnishings, cleansers or personal care products; and excess moisture. The best way to improve indoor air quality is to reduce pollutants at their source. Discourage smoking in the house, clean up any mold, and inventory and minimize chemicals that are stored and used in the house. These are only a few of the source reduction methods. Ventilation is a good secondary strategy that will help to dilute the pollutants in the air.

- Although ventilation can help to reduce humidity in the house during much of the year, there are times when it can actually make humidity problems worse. When it's hot and humid outside, the air may be almost saturated with water vapour. Because hot air can hold more moisture in suspension than cold air, bringing in hot, humid air to a cooler basement may result in drops of water condensing on cold surfaces such as basement floors and walls. When the weather is hot and humid, it's best to keep basement windows closed, ventilate only as required for fresh air for the occupants, and use a portable dehumidifier.

Safety

- Anytime you're working on electrical ventilation equipment, make sure the power supply to the unit is shut off.

Tasks

Clean bathroom and kitchen fans

An ideal ventilation system uses an HRV that continuously supplies fresh air and exhausts stale air. However, bathroom and kitchen fans can be an effective option. For some homeowners, particularly in isolated areas, the simplicity of kitchen and bathroom fans can be an asset. However, fans create static electricity, which attracts dirt to the fan and housing. For unimpeded airflow, you should keep bathroom fans clean.

Skill level rating: 1 - Simple maintenance

Materials: none

Tools: vacuum cleaner, rag, small brush

1. Pull down the grille. It's usually attached with two bent wire clips that can be slid out of their holes.
2. Unplug and remove the fan module, if possible.
3. Carefully brush, wipe or vacuum the fan blades and housing.
4. Wipe or wash the plastic grille. Allow it to dry thoroughly.
5. Re-assemble the fan and grill.

Clean range hood filters

There are two types of range hoods—vented that exhausts directly outside and unvented, that recirculate air inside the home. To ventilate a house, kitchen range hoods must vent air, moisture and odours directly outside. Recirculating range hoods depend on filters to capture some odours and grease. These filters are usually made of carbon that must be replaced frequently because of grease buildup. Both types of range hoods also usually have washable, aluminum mesh grease filters.

Skill level rating: 1 - Simple maintenance

Materials: none

Tools: dishwashing detergent and water, rag, vacuum cleaner

1. Follow manufacturer's instructions, if available. Remove the mesh filter.
2. Wash the mesh filter in a sink of soapy water. Rinse and let dry. (Note: Some filters can be cleaned in a dishwasher—check manufacturer's instructions).
3. Wipe the range hood housing with a damp rag.
4. Replace the filter.

Clean air intakes and exhausts to ensure clear airflow

Houses may have a fresh air supply to the return air duct of the furnace or a fresh air supply to an HRV. Exhaust fans and dryer vents should have an outlet on the exterior of the house. At the hood where they enter the house, fresh air supplies or stale air exhausts should have a 6 mm (1/4 in.) mesh rodent screen, not a fine bug screen. Dryer vents should have a properly closing flapper with no screen (lint will clog the screen too quickly). Intake bug screens quickly get clogged with dust and insects, making the intake useless. Stale air exhaust screens may become clogged with household lint or dust. Rodent screens do not clog as quickly but still need to be cleaned as part of seasonal maintenance.

If you can't find an exhaust hood for each bathroom fan, try to trace where each exhaust duct goes. Ducts that terminate in ceiling cavities or attics may contribute to mold and deterioration of the building.

Skill level rating: 1 - Simple maintenance

Materials: none

Tools: rag or brush

1. Locate the fresh air intake and exhaust hoods on the exterior of the house.
2. Remove leaves or any other obstructions that may be blocking the vents. During winter, clear any snow or frost buildup.
3. Unclip the grille (if necessary) and slide out the rodent screen.
4. Brush or wipe the screen. The exhaust vents also have a damper that may need to be cleaned so that it opens and closes properly.
5. Replace the damper (if necessary) and screen.
6. With the HRV, bathroom fan or furnace fan operating, perform the CMHC garbage bag airflow test (see page 62) to ensure that there is some airflow. There is no set requirement for the fresh air supply to a return air duct, but you do want to know if there's any flow at all.
7. At least once a year, remove the grilles and vacuum inside the ducts, as much as possible.

Replace inefficient bathroom fan and dryer duct systems

It is important to have air flowing smoothly and easily through ducts. Air will flow much more easily if the duct length, bends, kinks and other obstructions are minimized. Many bathroom fans and dryers, in particular, are installed with long flexible, plastic ducts that have corrugations to provide rigidity and bends or kinks where the ducts are hung. This arrangement results in poor airflow. Also, many dryers are vented inside basements. This adds excessive moisture to what are usually already damp spaces. Chemicals in detergents or fabric softeners also escape into the air.

Bathroom fans are often vented into ceiling spaces or attics. Again, this just adds moisture, often resulting in water condensing in those cool spaces. Adding moisture to hidden ceiling, wall or attic spaces can seriously harm the building.

The following instructions apply to ducts that terminate in a hood on the exterior of the building. For ducts that terminate inside an attic or basement, an appropriately sized hole must be created to continue the duct to the exterior of the house. This may require professional assistance.

Skill level rating: 2 - Handy homeowner

Materials: metal duct sections and fittings, self-tapping sheet metal screws, aluminum foil duct tape, plastic or metal duct straps

Tools: cordless drill, screwdriver

1. Inspect the dryer and bathroom fan ducts for duct material, kinks and bends.
2. If a shorter, straighter route from the appliance to termination on the outside of the building is possible, turn off and unplug the appliance. Many bathroom fans have a plug in the fan housing.
3. Remove the existing duct.
4. Plan the route for the new duct. Install the new duct, keeping bends to a minimum. A short section of flexible duct may be required at the dryer to enable it to be moved.
5. Make sure that the duct is hung securely and that all sections are screwed together with three screws.
6. Tape the joints between duct sections and where the duct attaches to the appliance and the exterior termination.
7. Make sure that the exterior hoods are in good condition and that the damper swings freely and closes properly. Replace any hoods that are in poor condition.
8. Make sure that any exhaust ducts are insulated if they pass through unconditioned space such as an attic. The water vapour being exhausted through dryer or bathroom exhaust ducts may condense into drops of water while travelling through un-insulated ducts that pass through a cold space. Liquid water inside the duct may hamper airflow or may flow back into the house if the duct is not sloped to the exterior. Exhaust ducts should be insulated to not less than RSI 0.5 (R3).
9. Restore power and turn on the appliance.
10. Perform the CMHC garbage bag airflow test (see page 62), either at the bathroom fan or at the dryer exhaust hood on the exterior of the building to ensure that air flows freely.

Inspect and clean heat recovery ventilators

Clean HRV filter, core and fans

Inside most HRVs are two air filters and a heat recovery core. The air filters are usually aluminum mesh while the core is commonly polypropylene. The following are generic instructions that usually apply, but it is always best to follow the manufacturer's instructions.

Skill level rating: 1 - Simple maintenance

Materials: soap and water

Tools: vacuum, garden hose, small brush

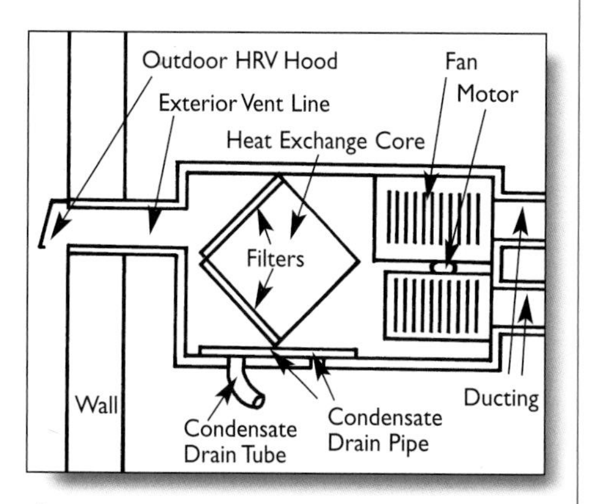

1. Turn off the HRV and unplug it.
2. Slide out the filters and heat recovery core.
3. Vacuum the filters; then wash them with mild soap and water.
4. Vacuum the heat recovery core. Either wash it with mild soap and water or spray it using a garden hose.
5. While the core and filters are removed, gently brush the fan blades to remove the accumulated dirt. Use a vacuum to remove the dislodged dirt.
6. Re-assemble the unit.

Clean the HRV condensate drain

In the HRV during the heating season, warm, moist air being exhausted from the house bypasses cold fresh air being drawn in. As the exhausted air cools, it cannot hold as much water vapour in suspension. Drops of water condense inside the HRV. The water must be able to drain through the plastic pipe or tube coming out of the bottom of the HRV. Because this route is really a plumbing drain, it must have a trap, which is usually in the form of a loop in the tube. Inspecting the condensate drain is easiest when the filters and core are removed for cleaning.

Skill level rating: 1 - Simple maintenance

Materials: 2 litres of warm, clean water

Tools: none

1. Turn off the HRV and unplug it.
2. Slowly pour about two litres (2 quarts) of warm, clean water into the drain inside the HRV.
3. If there's a backup, clean the drain by picking out any dead insects or other debris.
4. Make sure the tube is looped and isn't kinked or obstructed.
5. Repeat step 2 to ensure that the drain is flowing freely.

Seal HRV duct vapour barriers

The fresh air supply duct and the stale air exhaust duct that lead from the HRV to the outside of the building are both carrying cold air through the warm house. Water vapour from the house air may condense on any cold surface. That's why both of those ducts must be insulated. That insulation must be contained within a vapour barrier that is sealed to the HRV and to the wall where the ducts exit the building. If water vapour penetrates the vapour barrier, it may result in wet insulation or even chunks of ice around the exits. At the ports where the ducts are attached, most HRVs have an inner ring for the duct and an outer ring for the vapour barrier.

Skill level rating: 2 - Handy homeowner

Materials: Aluminum foil duct tape

Tools: utility knife

1. Inspect the ducts leading from the HRV to the exterior of the house. Look for sealed attachment to the HRV and to the wall. Feel along the length of the duct to ensure that the insulation coverage is complete. Look for tears or holes in the vapour barrier along the length of the duct. Look for spots where water has dripped. In winter, feel for chunks of ice in the insulation inside the vapour barrier.
2. Seal any minor holes or tears with aluminum foil duct tape.
3. If the insulation is dislodged at each end, loosen the vapour barrier and slide the insulation to provide complete coverage.
4. Seal the vapour barrier at each end (to the outer ring, if it exists) with aluminum foil duct tape.
5. If there are major holes in the vapour barrier or gaps in the insulation, the duct may have to be removed and reinstalled. You may choose to do this yourself or seek professional assistance.

WINDOWS AND DOORS

Windows and doors provide ventilation, light, access and security to your home. Repairs tend to fall into two areas—poor operation and breakage. A badly operating or broken window or door can be dangerous in an emergency, be an invitation to an intruder and cause heat loss that will increase your energy costs.

The most common window and door problems are:

- water leaks around windows, doors, skylights or other penetrations
- damaged window and door weatherstripping
- window condensation
- broken windowpanes
- damaged screens
- problem lock sets or interior passage sets
- sticking doors (interior and exterior)
- self-closure devices
- garage overhead door malfunctions

Maintenance includes:

- regular inspection to spot problems before they become serious

Prevention tips

- Keep window tracks clean and lubricated for easy operation.
- Repair minor problems promptly to avoid bigger repairs. For example, door stops that come loose or go missing need to be reinstalled right away to avoid damage to walls by the doorknob, a bigger patching job.

Repair tips

- When repairing a broken pane of glass in a storm door or window, the job will be easier if you remove the door or window and work on a flat surface.
- When replacing a screen, use a work table or floor space big enough to move around the job.
- A staple gun is easier to use when attaching screen to the frame, but a hammer and tacks will still do the job!
- Purchase replacement glass, screen and screen replacement kits at your local hardware or building supply store.
- A spline roller is the best tool to use when replacing the spline in a metal or vinyl frame screen. A screwdriver can be used as a substitute, but you may need a second person to help keep the screen tight in the frame.

Special considerations

Healthy Housing™

- Regular maintenance for weathersealing of windows and doors helps to control the heating and cooling efficiency in your home. This sealing helps to keep your home more comfortable and saves money on your heating and cooling bills.
- Whenever possible, choose low-toxicity, solvent-free caulking that has less affect on indoor air quality and your health.

Safety

- Protect your eyes and hands when removing broken glass or installing new glass in a frame. Wear protective eyewear and gloves.
- Replace broken glass right away to prevent someone from getting injured.

Tasks

Repair water leaks around windows, doors, skylights and other penetrations

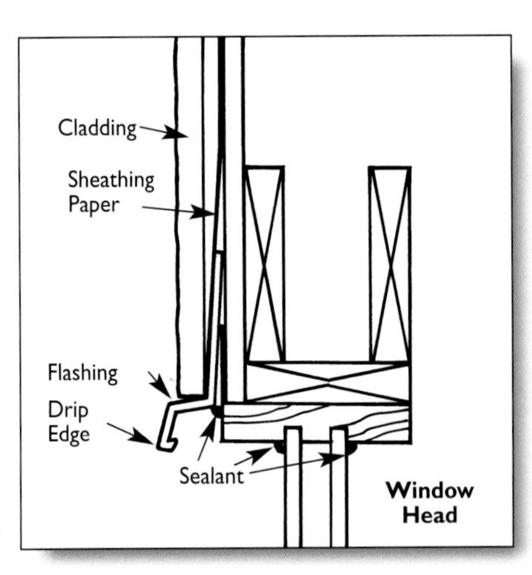

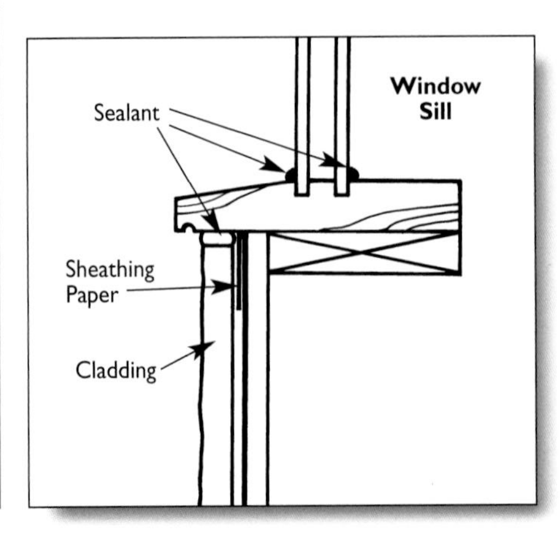

Poor flashing or deteriorated caulking can allow water to get into the walls around window and door frames. Ideally, drip caps would have been installed over windows and doors that are not well protected by a roof overhang. Flashing kits around skylights are designed to prevent water from penetrating with only minimal use of caulking. Sealants do not last forever and must be maintained.

Many different sealants are suitable for repairing water leaks on the exterior of the house. Siliconized acrylic sealant combines the durability of silicone with the ease of use and paintability of acrylics. Read the label closely to make sure that the caulking material is compatible with the surfaces where you intend to apply it and that it can be used inside. Provide ventilation when caulking because many compounds release potentially harmful volatile organic compounds while curing. Avoid using exterior products inside where the strong odours will last for days or weeks and may affect your health.

Skill level rating: 2 - Handy homeowner

Materials: metal or vinyl flashing, siliconized acrylic or acrylic sealant, contractor's sheathing tape

Tools: caulking gun, water, hammer

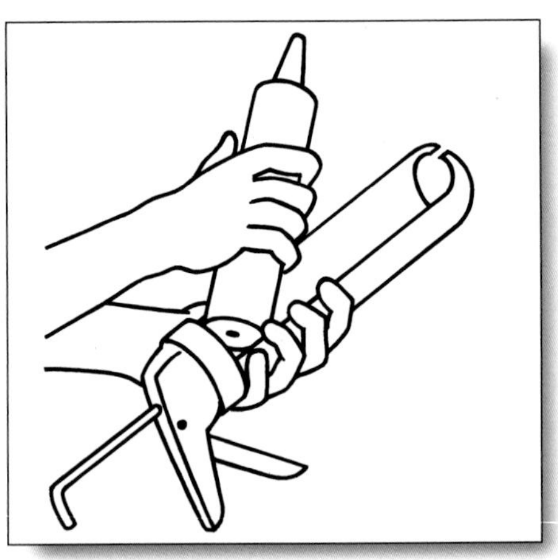

1. Use a rag and putty knife to gently remove any old caulking or dirt from a gap to be caulked.

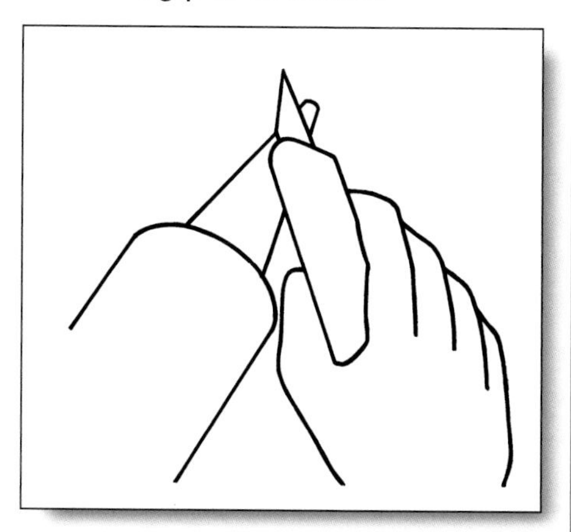

2. Cut the nozzle of the caulking tube on a slight angle. The size of the cut opening, the pressure you apply to the gun and the speed at which you lay out the bead, will determine the width and depth of the caulking bead.

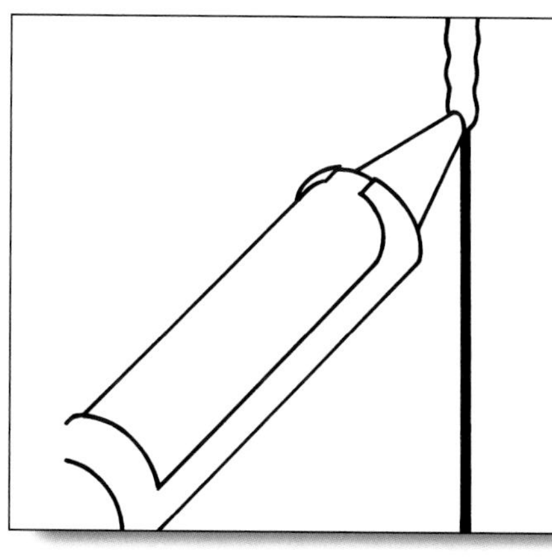

3. Apply the caulking by squeezing the trigger on the gun while moving the nozzle slowly and smoothly along the joint.
4. Quickly use a wet finger to smooth the caulking before it skins over.

Replace window and door weatherstripping

Weatherstripping is an important part of the house that keeps heat in and wind and water outside. Water leakage can cause mold and damage to the structure and interior finish of the house. Air leakage can waste energy and cause uncomfortable drafts. As part of your seasonal home maintenance inspection, check the condition of all window and door weatherstripping. Because windows and doors are subject to many openings and closings, weatherstripping can wear out.

There are a number of different types of weatherstripping and door sweeps. Many are designed to be compressed in an opening to provide a seal. In some windows or doors, compression strips may be designed to lock into a groove or kerf, avoiding the need for surface fastening. Low-cost metal spring type and foam weatherstripping may not prove to be very durable. Metal clad doors may also have a magnetic weatherstrip system that is attracted to the metal surface of the door. Modern windows and doors often have two lines of weatherstripping of slightly different types.

Skill level rating: 2 - Handy homeowner

Materials: Weatherstripping kit (similar to the original weatherstrip if possible)

Tools: measuring tape, utility knife, hacksaw, screwdriver, hammer

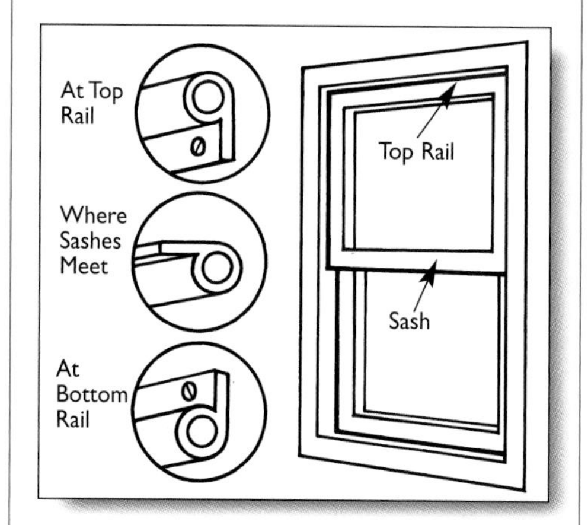

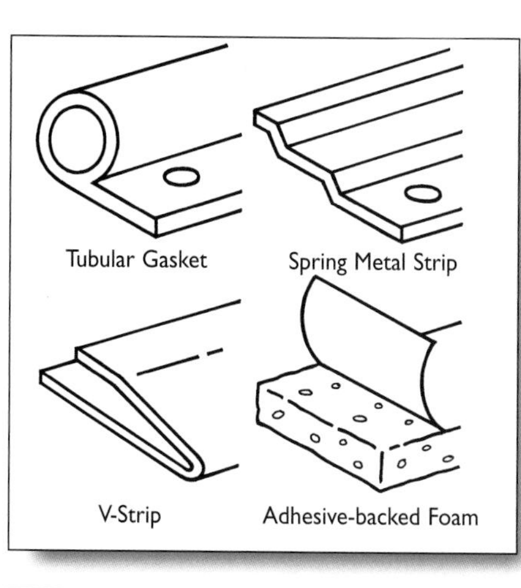

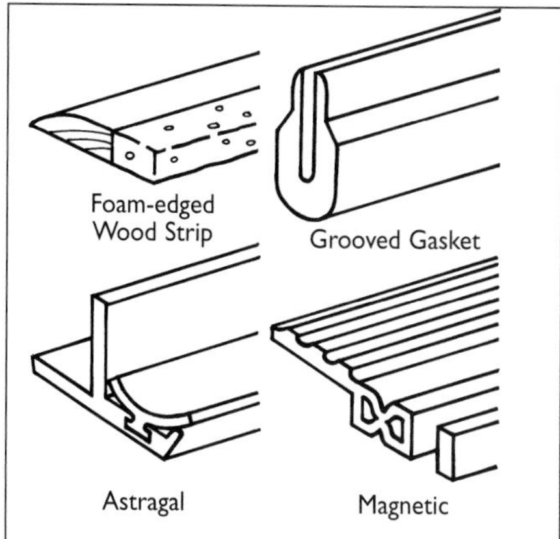

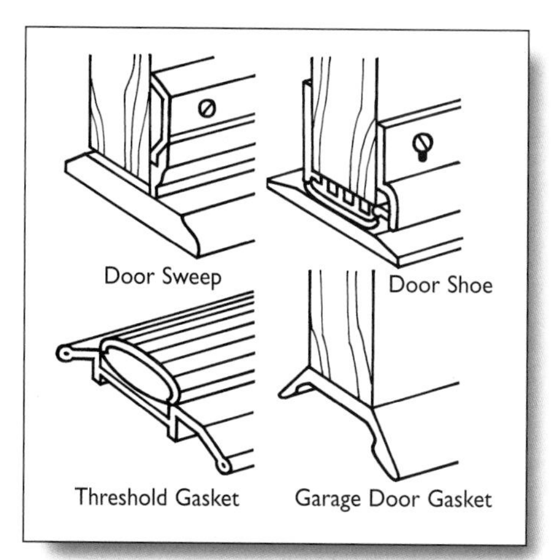

1. Remove the worn weatherstripping.
2. Measure the old weatherstripping or the opening.
3. Cut the new weatherstripping to fit.
4. Install the new weatherstripping (as appropriate for the type).
5. Check the window or door operation and tightness of the seal. Adjust weatherstripping if required (depending on type, it may have to be removed and placed again further from window or door).
6. Follow the same procedure for a door sweep.

Fix problem windows

Loose windows usually have broken or improperly adjusted operating mechanisms. Older windows use weights attached to sash cords or chains to operate. Newer windows use springs and can be adjusted with a screwdriver. Sticking windows commonly either have dirty, unlubricated sliders or have been painted shut.

Skill level rating: 3 - Skilled homeowner

Materials: sash cord (depends on window type), graphite- or silicone-based lubricant

Tools: utility knife, pliers, screwdriver, hammer, small pry bar (optional), toothbrush or small brush, cloth

Fix sticking window

1. Check to see if the window has been painted shut. If it has, use a utility knife to cut the paint seal or place a small block of wood against the window sash and tap gently with a hammer to free the window.
2. Clean any debris or dirt from the window tracks using a brush and cloth.
3. Lubricate tracks using a graphite- or silicone-based lubricant.

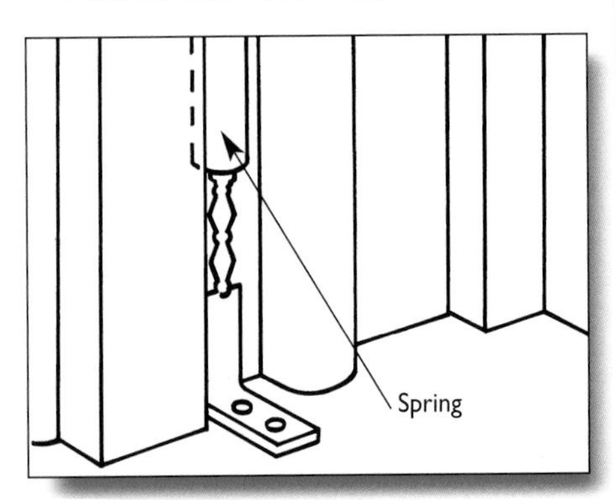

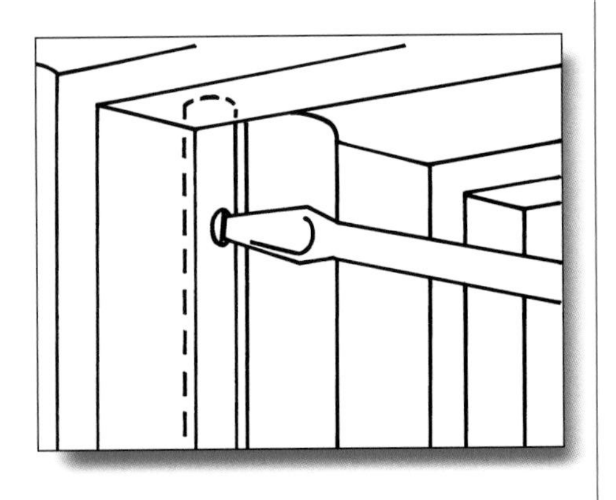

Adjust spring-loaded windows

1. Locate the adjustment screw on the window jamb.
2. Turn screw to adjust the window balance. Test the window for smooth operation.
3. Continue to adjust the screw and test the window until it works smoothly.

Replace broken sash cords with weights or chains

1. Pry away and remove window stops from frame using a screwdriver or pry bar. Remove moulding screws if present.
2. Remove any weatherstripping or moulding if present.
3. Gently remove lower window. Pull sash cords or chains out of window sashes.

4. Pry away weight channel cover and remove the weight from inside the window frame. Detach old sash cord from the weight.

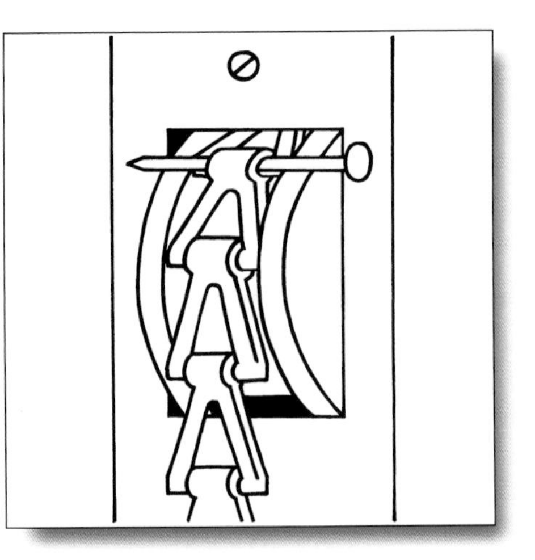

5. Install new sash cord or chain. Thread new cord or chain over pulley in upper window frame and drop into the channel for the weight. Pull on new cord or chain at both top and bottom to make sure it runs smoothly over the pulley.
   Note: To make it easier to thread the new sash cord or chain over the pulley and down through the weight channel, attach a string with a nail to the end of the sash cord. Thread the nail and string over the pulley and let it drop down into the channel, carrying the sash cord. Grab the sash cord or chain, remove the string and nail, and continue.

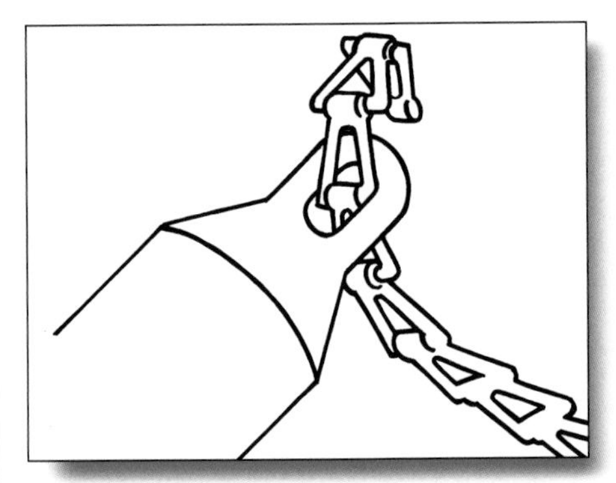

6. Attach the new sash cord or chain to the weight and place in the channel. Pull on sash cord or chain above the pulley until the weight is against the pulley.
7. Sit window on sill. Holding sash cord or chain against the window, cut the cord or chain 75 mm (3 in.) longer than the hole in the window sash.

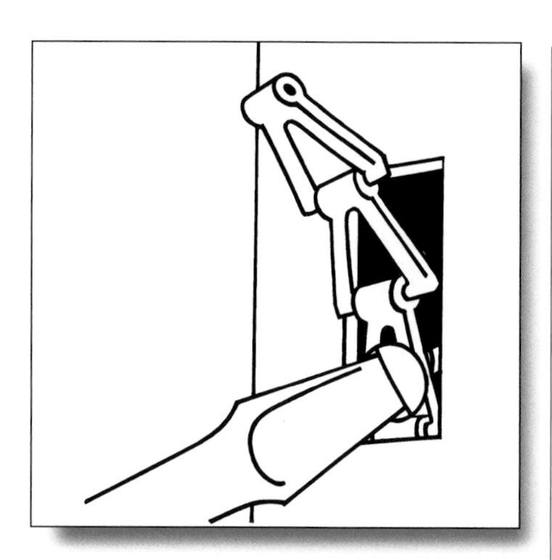

8. Tie a knot in the sash cord and wedge it into the hole in the sash. If using chain, attach the chain to the weights with a wire. Snip off excess chain. Attach chain to the sash with a screw.

9. Replace the weight channel cover. Slide the window into the frame. Reattach stops and any moulding or weatherstripping that had been removed.

Replace broken windowpanes

A broken glass pane should be replaced immediately. It can injure someone, let in rain and snow, cause condensation and let precious heated air escape in winter. The moisture can quickly cause sill, floor, moulding and wall damage. Cold drafts and loss of heat will soon make the house uncomfortable. Broken panes can also invite vandals and burglars looking for an easy way to enter.

Windows with sealed glass units cannot be repaired. Order replacements from a manufacturer. If you have a warranty, check to see if repairs are covered. Use manufacturer's recommendations for replacement of sealed units.

Older windows with wood or metal frames can be repaired at home.

Skill level rating: 2 - Handy homeowner

Materials: putty, glazier's points, masking tape

Tools: gloves, eye protection, pliers, replacement glass, putty knife, small chisel, screwdriver, paintbrush

Wood frame windows and storm doors

CAUTION: Wear gloves and eye protection.

1. Wearing gloves and eye protection, gently remove the pieces of the broken pane.
2. Remove leftover pieces of glass embedded in the frame by rocking them back and forth while pulling them away from the frame with pliers.
3. Pull out the old glazier points (or nails used for the purpose) with the pliers.
4. Measure the height and width of the cleaned frame. Subtract 5 mm (3/16 in.) from each dimension and have a new piece of glass cut to this size at your local hardware, building supply or glass replacement store.
5. Knead putty so it is uniform and press a thin putty ribbon around the inside edge of the frame.
6. Place the new glass in the frame and press it against the putty firmly and evenly.
7. Secure the glass with new glazier points placed at the corners and every 100 to 150 mm (4 to 6 in.) along the frame. To avoid breaking the glass, gently push or tap the points in with the side of the chisel or a screwdriver. Do not be tempted to use a hammer!
8. Fill the joint with putty or glazing compound, pressing it firmly and smoothly into place with your fingers. Use a putty knife to cut away any extra putty or compound.
9. After the putty has dried a few days, paint it to match the rest of the window. Clean the window after the paint dries.

Metal and vinyl frame windows and storm doors

Many modern windows, window units and storm doors have metal or vinyl frames. Although a window may be an exterior slider or storm, it's frequently designed to be removed and repaired from the inside of your home.

Repairs depend on the type of frame and how it holds the glass in place. The most common types of frames used in residential construction are:

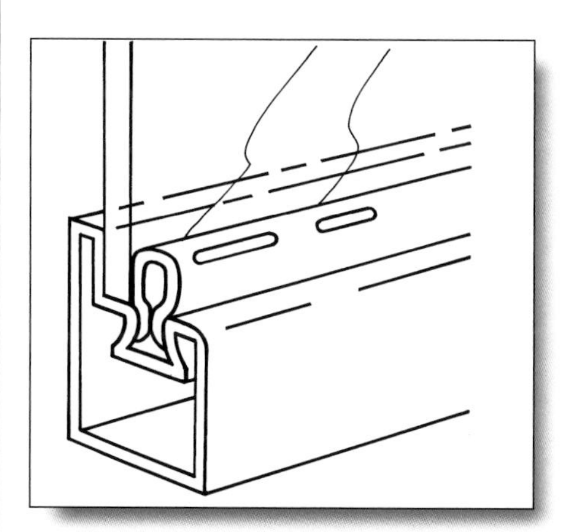

- The sealed glazing unit fits against the vinyl or wood sash like a picture in a frame and is held on the interior with vinyl or wood strips. A rubber seal holds the glass in place.

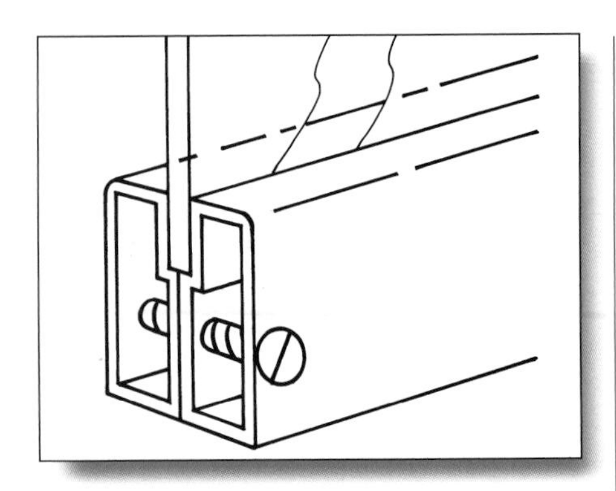

- For storm windows or doors, the glass may rest in a channel in the frame. Screws or metal keys hold the frame together. Putty or some form of rubber seal holds the glass in place.

Alternatively for storm windows or doors, the glass may be sandwiched between two separate frame halves that are held together by rivets or screws. A rubber or vinyl seal, attached along each inside half of the frame, makes the joint airtight.

Materials: EITHER masking tape, putty or mastic compound, window glass cut to fit the frame opening, rubber or vinyl seal, matching paint, OR replacement glass and sash unit

Tools: measuring tape, screwdriver, hammer, wood chisel, pliers, putty knife, paintbrush

Replace a sealed glazing unit

1. Check carefully along the edges of the window to discover the manufacturer's name.
2. Measure the overall size of the glazing unit and the frame or sash in which it is contained. Contact a building supply store or distributor of that brand of windows and find out the replacement procedure.
3. For operable windows, manufacturers will often sell the complete glazing unit and sash so that replacement is just a matter of unfastening the window from the existing hardware and installing the new window in its place.
4. Follow the manufacturer's instructions.
5. Replacing glass in larger fixed windows will likely require professional help.

Replace glass in two frame halves

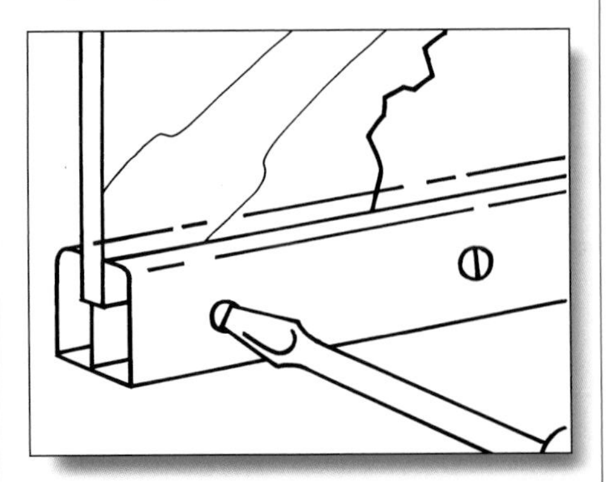

1. Remove the screws holding the two frame halves together; then remove one of the frame halves. Carefully remove the damaged glass.
2. Measure the inside height and width of the frame opening and subtract 3 mm (1/8 in.) from each dimension to allow for irregularities in the frame. Order the glass cut to size at your local hardware or building supply store.

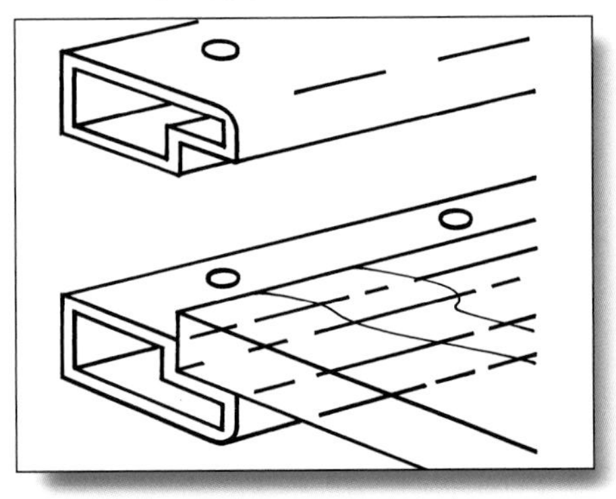

3. Place the new glass in the frame on the seal.
4. Screw the two halves back together.

Replace glass in channel frame

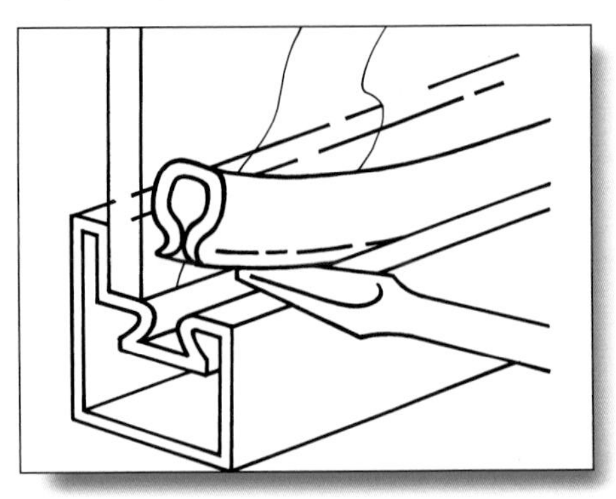

1. Remove the glazing seal from the frame or sash with the tip of a screwdriver and carefully remove the damaged glass. Open one end of the frame by gently knocking out the metal keys or removing the screws that hold it in place.

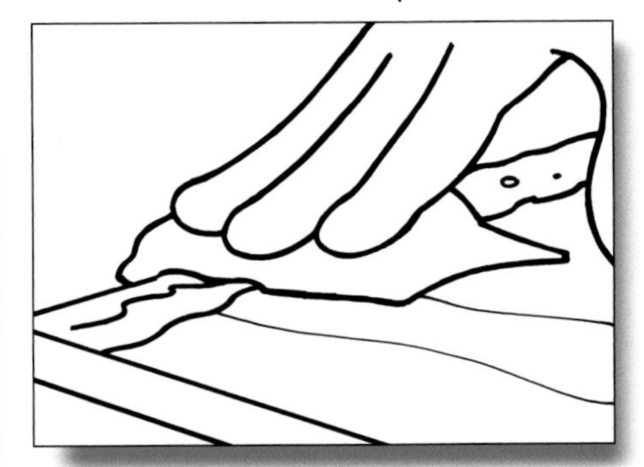

2. If mastic has also been used to seal the glass, thoroughly remove it from both the frame and the glazing seal with a cloth soaked in solvent. **CAUTION:** Use solvent only in a well-ventilated area.
3. Measure the inside height and width of the frame opening and subtract 3 mm (1/8 in.) from each dimension to allow for irregularities in the frame. Order the glass cut to size at your local hardware or building supply store.
4. Using the putty knife, spread a layer of new putty or mastic on the frame where the glass is to rest. Place the new glass in the opening and press it firmly against the mastic. Apply a 3 mm (1/8 in.) thick layer of putty or mastic along the edge of the glass equal in width to the raised portion of the glazing seal.
5. Replace the glazing seal. Examine its edges to be sure the mastic is tight between the glass and the bead. Remove any excess mastic along the edge of the bead with the putty knife, then use a cloth and solvent (with caution) to finish the cleaning.

Fix or replace screens

Screens with holes or tears allow insects to get into your house. You can easily fix or patch small holes or tears. If the tear is large or hard to patch, replace the entire screen. Old windows or screen doors often have metal screening. Newer windows usually have glass fibre screening.

Skill level rating: 2 - Handy homeowner

Materials: screening or ready-cut screen patches

Tools: old scissors, a ruler or small block of wood with straight edges, sewing needle and fine wire or nylon thread

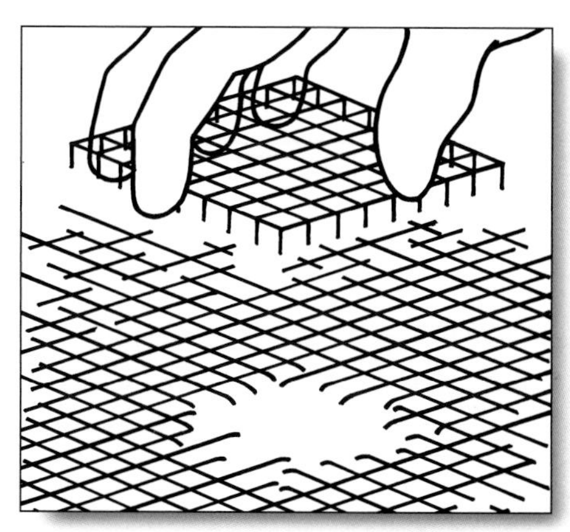

Patch a hole or tear in metal screen

Note: The patch should be put on from the inside surface of the screen so that it will look tidy from the inside when the job is finished.

1. Trim the hole or tear with the shears, cutters or old scissors to make a rectangle with square edges.
2. Cut a similarly shaped rectangle from your screening that is 25 mm (1 in.) larger on all sides than the hole.
3. Remove the three outer wires on all four sides of the patch.
4. Using the block of wood or ruler as a frame, bend over the edges of the wire.
5. Put the patch over the hole and push the bent wires through the screen.
6. Hold the patch firmly in place, then from the other side of the screen bend the loose wire ends in toward the hole. You may need someone's help to do this.

Patch a hole or tear in glass fibre screen

1. Trim the hole or tear with old scissors to make a rectangle with square edges.
2. Cut a similarly shaped rectangle from your screening that is 25 mm (1 in.) larger on all sides than the hole.
3. Pull "threads" from a piece of screen and use them to sew the patch into place.

Replace the entire screen

Wood frames

Skill level rating: 2 - Handy homeowner

Materials: screen fabric, screen moulding, finishing nails, staples or tacks

Tools: measuring tape, staple gun, utility knife, screwdriver, hammer, crosscut saw (if needed)

1. Measure the length and the width of the screen opening. Cut screening 150 mm (6 in.) longer and 75 mm (3 in.) wider than the opening.
2. Remove the door or window and place it on a flat surface. Using the screwdriver, gently pry up and remove the moulding that holds the screen in place. Remove the old screen.

3. Align the new screen over the frame. Place the screen about 25 mm (1 in.) above the top of the opening. Staple or tack the screen every 50 mm (about 2 in.) across the top of the frame.

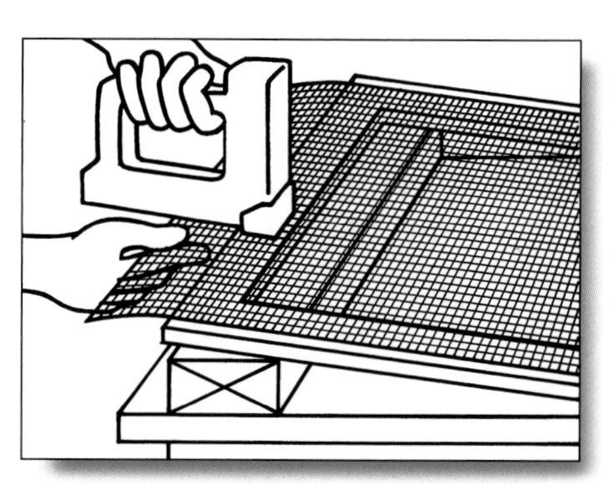

4. Pull the screen tight over the frame lengthwise. Tack or staple the screen every 50 mm (about 2 in.) across the bottom of the frame. Note: To help pull the screen tight you can tack or staple a board along the bottom of the screen overlapping the frame. Slip the bottom edge of the frame over the edge of your work surface. Place pressure against the board until the loose screen is tight against the frame. Hold the board in this position while you tack or staple the screening to the frame.

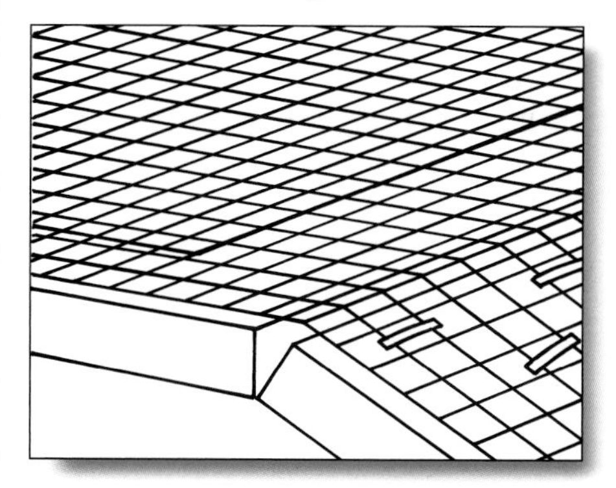

5. Staple or tack the screen every

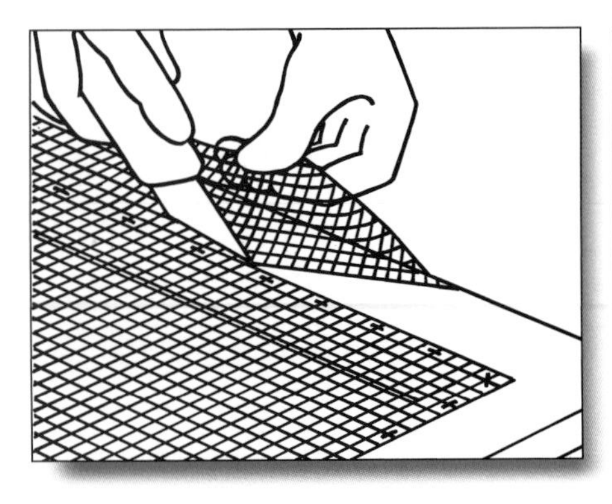

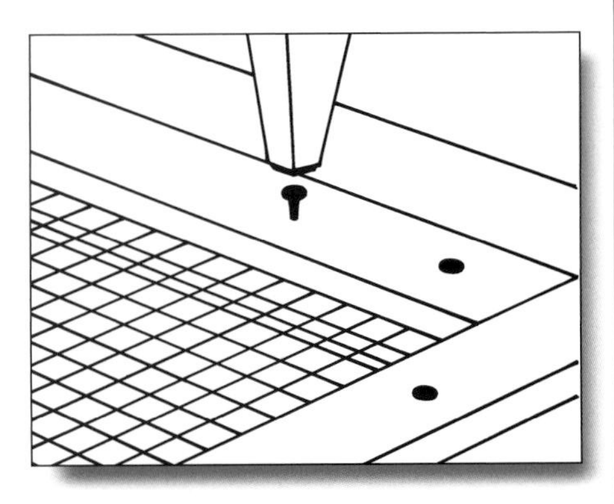

50 mm (about 2 in.) down the centre rail, if any. Pull the screen tight at the sides and attach to the frame. Cut away the excess screen and reattach the moulding using finishing nails.

Metal and vinyl frames

Skill level rating: 2 - Handy homeowner

Materials: screen fabric, spline

Tools: measuring tape, screwdriver, spline roller

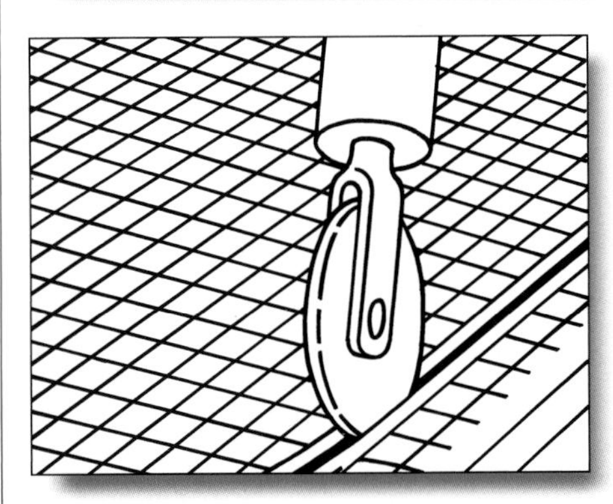

1. Measure the length and width of the screened opening. Cut the screening 150 mm (about 6 in.) longer and 75 mm (about 3 in.) wider than the opening.
2. Remove the window or door and place it on a flat work surface.
3. Remove the old screen. It is usually held in place by a vinyl spline, forced into a groove in the frame. Pry out the spline using a screwdriver.
4. Carefully align the new screen over the frame. Keep the screen tight and use a spline roller to work the spline back into the grooves.
5. Cut away the excess screen with the utility knife.

Replace problem lock sets or interior passage sets

Locksets or passage sets (that have no locking mechanism) are usually durable as long as they are securely fastened. Usually, there are machine screws through the inside rose (hole cover) that connect the inside and outside knobs through the latch bolt assembly. There are also two wood screws through the latch plate on the edge of the door. As part of seasonal maintenance, make sure that these screws are all secure. At the same time, ensure that hinge screws are secure. With the door closed, use a hammer and nail set to strike the hinge pins up, one at a time. Lubricate each pin with silicone lubricant and tap back into place.

Eventually, locksets or passage sets may wear out and need to be replaced.

Skill level rating: 2 - Handy homeowner

Materials: replacement lockset

Tools: screwdriver, hammer, chisel

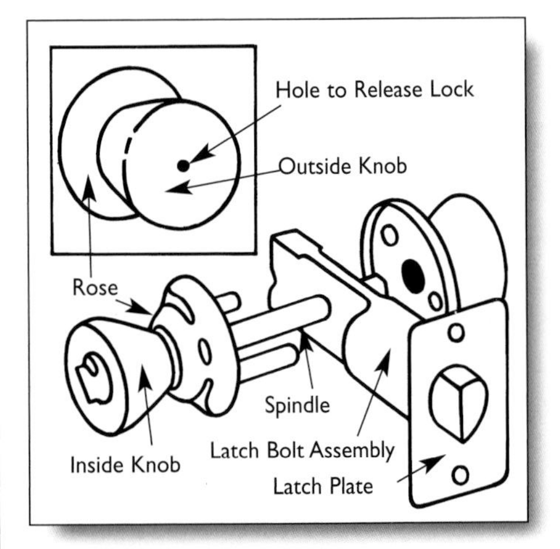

1. Measure the distance from the edge of the door to the centre of the base of the doorknob. This distance is the backset. For a more accurate measurement, temporarily remove the doorknobs and latch bolt assembly. Remove the two machine screws on the inside rose that hold the two doorknobs in place. (Some knobs have a release tab at their base; press a nail into the tab to release.) Remove the rose. Slide out the doorknobs and connecting spindle. Remove the two screws in the latch plate. Slide out the latch plate and latch bolt assembly.

Measure the distance from the latch plate to the centre of the spindle hole in the latch bolt assembly.

2. Choose a new lockset or passage set that has the same backset and will fit the holes in the door.
3. Insert the new latch bolt assembly and secure it with the two screws through the latch plate into the door. If necessary, use a chisel to make any minor adjustments to the mortise in the edge of the door.
4. Insert the spindle on the inside knob through the spindle hole in the latch bolt assembly.
5. Position the outside knob on the spindle and line up the screw holes.
6. Secure the knobs in place with the machine screws.
7. Make sure that the door closes and that the latch fits into the strike plate. If necessary, adjust the position of the strike plate and strike plate recess on the door frame.

Adjust sticking doors

(interior and exterior)

Doors may have a number of problems. They squeak, stick, drag and may not close properly because they keep striking the frame or because the lock catch and strike plate are misaligned. Doorknobs may rattle and hinges may be loose. Patio door tracks can become dirty or misaligned.

These problems are irritating but are usually fixed quite easily.

Skill level rating: 2 - Handy homeowner

Materials: graphite or silicone spray lubricant, oil, sandpaper, small dowels or toothpicks, putty

Tools: screwdrivers, hammer, plane, pliers

Fix noisy doors

1. You can easily stop a door from squeaking. Put a few drops of oil or silicone spray lubricant at the top of each hinge. Move the door back and forth to work the oil down into the hinge. If the squeaking does not stop, raise the pin and add more oil.
2. Lubricate noisy or squeaking locks with graphite or silicone spray lubricant, available at your hardware store.
3. To stop an older doorknob from rattling, loosen the set screw on the knob. Remove the knob and put a small piece of putty inside it. Push the knob back as far as possible and tighten the screw. New doorknobs may not be able to be tightened and will have to be replaced.

Fix sticking or dragging doors

Swinging doors

1. Check for loose screws in the hinges, and tighten them as necessary. If the screws won't hold, replace them one at a time with longer screws of the same gauge or insert a small dowel or toothpicks in the hole, break off the exposed ends, and put the original screw back.
2. If the door still sticks or drags, look for a shiny spot on its edge where the paint or finish is marked. Pinpoint the spot by opening and closing the door slowly several times. Sand the spot down until the door closes smoothly. Don't sand too much, or the door will not fit tightly as it should.
3. If the door frame is badly out of shape, you may have to remove the door and plane down the edges that drag.
4. If necessary, paint the frame to match.

Sliding patio doors

Skill level rating: 2 - Handy homeowner or (if track is badly damaged)

Skill level rating: 4 - Qualified tradesperson/contractor

Materials: silicone spray lubricant

Tools: vacuum, hammer

1. Clean out any dirt or debris in the tracks. A vacuum is handy for this job.
2. Once tracks are clean, lubricate with a silicone lubricant.

3. If the track is slightly damaged, try inserting a block of wood and hammering the damaged track flat.
4. If the damage is severe, you will need to replace the track. This job is often one for a qualified patio door installer. If you decide to do it yourself, order a track replacement kit from the door manufacturer. Follow the manufacturer's instructions to replace the track.

Replace door hinges

Door hinges can wear out, rust and become loose, causing the door to be out of alignment. Doors that are exposed to the weather are more likely to have damaged or badly rusted hinges that need to be replaced. There are two main types of door hinges—loose-pin and fixed-pin. Loose-pin butt hinges are the most common. They have a removable pin that allows you to remove the door without unscrewing the hinge leaves. Fixed-pin hinges must have the hinge leaves unscrewed.

Remove loose-pin hinges

1. Remove the pins from the top and bottom hinges by tapping the pins upward first with a hammer and nail set through the bottom hole in the barrel and then with a hammer and screwdriver under the edge of the head of the pin. If you are doing this job yourself, it's usually easier and safer to remove the bottom pin first.
2. Stand the door on its edge on the floor and unscrew the hinge leaves attached to it. Unscrew the hinge leaves from the door frame.

Remove fixed-pin hinges

1. Remove the door by first removing the screws that attach the hinge leaves to the door. Undo the bottom hinge first.
2. Stand the door on its edge on the floor and unscrew the hinges from the door frame.

Replace the hinges and reattach the door

1. Pack any enlarged screw holes with wood filler and let dry. Drill new pilot holes for the hinge screws.
2. Attach the new hinges or appropriate hinge leaves on the door in the same position as the old ones. Tighten the screws one-half turn short of tight.
3. Screw the opposing leaves of loose-pin hinges onto the door frame.
4. Use small pieces of wood under the door or have a helper hold the door high enough to position the door so that the two halves of the hinges mesh together. Insert the hinge pins.
5. Tighten all screws and check the door for close and fit. If it binds or rubs, loosen the screws and slightly shift one or more of the hinges. Small pieces of cardboard cut from the box that the hinges were packaged in can be used to shim the leaves of the hinges either on the door edge or the door frame.

Adjust self-closure devices

A self-closure device is very important on a door between the house and an attached garage. It ensures that the door stays closed. This closure minimizes the chance that deadly carbon monoxide from idling vehicles may enter the house. Self-closure devices are also used on screen doors to ensure that they close without slamming.

Skill level rating: 2 - Handy homeowner

Materials: none

Tools: screwdriver

1. Ensure that all self-closure device mounting screws are secure. Tighten as required.
2. Open the door and allow it to close. Ensure that it closes completely but not with excessive force.
3. If the door does not close properly, check the fit of the door. Make any necessary adjustments by shimming out hinges or adjusting the lockset strike plate.
4. Most self-closure devices have an adjustment screw on the body of the closer. Turn the screw until the closing pressure is acceptable.

Maintain garage overhead doors

Sectional garage doors have many moving parts that need semi-annual lubrication. Springs may also need adjustment or replacement.

Skill level rating: 2 - Handy homeowner

Materials: silicone lubricant

Tools: rag, screwdriver, wrenches

1. Check all hinges and brackets to ensure that all screws and bolts are secure.
2. Wipe the track clean with a rag.
3. Lubricate the track and rollers with silicone lubricant.
4. Check that the door operates smoothly and that all lock bars move easily into the slots in the track. Adjust lock bar brackets as required.
5. Lubricate the door lock with a lock lubricant or powdered graphite.
6. Check the balance of door. It should not seem heavy to lift, nor so light that it lifts on its' own. If either of these conditions exist, the springs must be adjusted or replaced.
   Warning: Springs are under high tension and should only be adjusted or replaced by a qualified person.

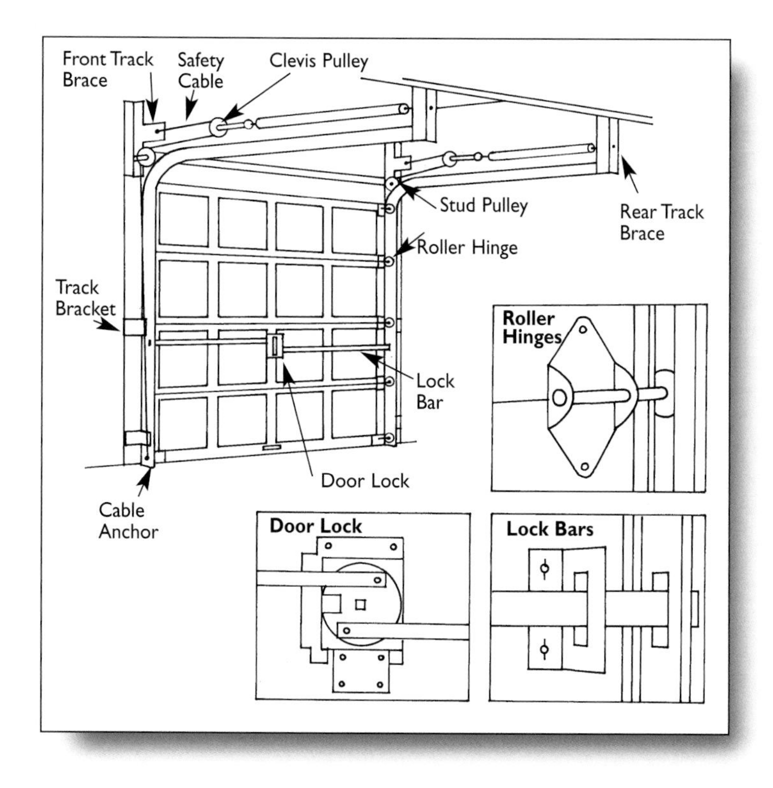

SIDING

The most common siding problems are:

- poor appearance
- buckling or detachment
- deteriorated caulking or flashing that may allow water penetration

Maintenance includes:

- cleaning, re-fastening and making minor repairs on vinyl and aluminum siding
- painting and staining wood siding
- minor re-pointing of masonry
- inspecting, reattaching and caulking flashings and any gaps

Prevention tips

- Protect your investment. Conduct regular (one to two times a year or after a windstorm) inspections of your home's exterior to spot trouble early and prevent severe problems. Look for signs and locations of potential moisture problems including cracks, gaps, leaks, obvious deterioration, staining, material warping and efflorescence on brick. Paint bubbling, cracking and peeling can indicate an underlying problem, such as moisture movement from inside (air leakage/condensation).
- Include in your regular inspections soffits, fascia, eavestroughing, downspouts and window and door caulking. These elements are important parts of the exterior finish system.
- Maintain all elements of the exterior siding, making general repairs and applying paints, coatings or other preventive measures as recommended by the manufacturer.
- Keep soil at least 200 mm (8 in.) below the exterior finish to prevent moisture damage. It is acceptable to allow soil to be only 150 mm (6 in.) below brick, although 200 mm (8 in.) is preferred.

Repair tips

- When preparing to paint, take the time to check and repair minor damage that could lead to bigger repair problems.
- Most caulking is limited to bridging about 60 mm (0.25 in.) between solid materials. Fill large cavities and holes with fiberglass, polystyrene rope or spray foam. Cover with a solid material such as wood or aluminum flashing so that only small gaps remain. Caulk remaining gaps.
- Rent and use scaffolding to provide a safe, working platform.

Special considerations

Healthy Housing™

- Effective air sealing will help reduce energy consumption and moisture and mold problems.

Safety

- Siding repairs and maintenance often require climbing on ladders. Know how to use a ladder safely. Consider renting scaffolding and safety equipment to do repairs.
- If your home is more than 40 years old, you should assume that the paint on your home contains lead. This fact is not always a cause for alarm. Lead-based paint is not dangerous if it is in good condition, but if it is peeling and flaking then the paint presents a potentially harmful situation. Sanding and scraping lead-based paint can also produce large amounts of dust that contains lead.
- Those especially at risk from lead-based paints are infants, young children, pregnant women and the fetus. Paint samples can be tested with a home test kit or through laboratory analysis. Current federal and provincial laws restrict the amount of lead that can be contained in commercial products. New paints do not contain lead.

Tasks

Maintain vinyl and aluminum siding

Vinyl and aluminum siding are durable and low maintenance. Regular maintenance includes inspecting to find any damage, cleaning, caulking cracks or holes, and ensuring that soil is kept below the siding to prevent moisture damage to the house. In unusual situations where a piece of the siding is severely damaged (large dents, holes, cracks or breaks), the affected piece may need to be replaced. Consult with a qualified siding contractor.

Skill level rating: 2 - Handy homeowner

Materials: non-abrasive soap, water, caulking

Tools: long handled brush, hose, caulking gun, rake or shovel

1. Clean surface stains gently with water and a non-detergent, non-abrasive soap. Start cleaning at the bottom and work up to avoid streaks. You can use a long handled brush to clean the siding, then hose gently (high pressure could force water behind the siding and flashings which are designed to shed water coming down, not upwards as from a pressure washer spraying from the ground up).
2. Caulk any visible cracks and replace old caulking to keep gaps well sealed.
3. Keep soil 200 mm (8 in.) below the lower edge of the siding to prevent moisture damage to the house structure. Be sure to maintain the grade so that it slopes away from the house.

Maintain wood and wood-based siding

It's important to keep any type of wood or wood-based siding dry to prevent deterioration and mold growth. Regular maintenance includes inspecting to find cracks or moisture problems, repairing any damaged pieces, cleaning and painting.

Skill level rating: 2 - Handy homeowner

Materials: non-abrasive soap, water, caulking

Tools: long-handled brush, hose, caulking gun, rake or shovel

1. Scrub wood siding lightly using a mild, non-abrasive and non-detergent soap and a long brush. Rinse by hand or hose off the soap (high pressure may force water through the siding).
2. Premature deterioration such as cracks or discoloration can be caused by leaks. Check caulking and replace where necessary. Replace old, cracked caulking.
3. Check for water pooling as a result of poor grading and drainage or splashing from the eavestrough. Prevent splashing by installing splash blocks, available at a building supply or hardware store.
4. A piece of damaged or deformed siding can be removed and replaced. Consult a renovator or contractor for assistance.

Paint wood siding

Painting your exterior siding will help protect your home from moisture and weather damage. Painting should be done when you begin to see signs that the paint is wearing or peeling. As paint wears, the primer begins to show through the finish coats, the color of the old paint dulls and the grain on wood siding begins to show clearly.

Wait until you see these signs. If you repaint too often you will end up with a thick coating that's likely to crack and break away. The coating will be so thick that won't be able to expand and contract with changes in the weather or adjust to your home's natural structural shifts. If you wait too long, the paint surface deteriorates making it hard to restore. In that case, you may need to sand down to the bare surface and begin again. By regularly checking the condition of the exterior paint, you will be able to judge the right time to repaint.

Skill level rating: 2 - Handy homeowner

Materials: paint, paint primer for wood

Tools: wire brush or steel wool, rollers and tray, paint brushes 50 mm (2 in.) to 100 mm (4 in.) wide, preferably with chisel edges, an angular sash brush (good if you are painting window frames, mouldings or other narrow surfaces), detergent, sponges or cleaning cloths, drop cloths, paint bucket, mixing paddles, protective clothing, safety equipment, an extension ladder, step ladder, sander or heat gun

Before you paint or repaint, check the surface carefully and make repairs.

1. Replace decayed or damaged wood around windows, steps, posts, eaves and anywhere else.
2. Clean eavestroughs and downspouts before painting.
3. Use a wire brush and steel wool to clean rust from all metal surfaces. Paint these surfaces with a metal primer before applying the finish coat.
4. Use drop cloths to protect foliage and windows against paint splatters.
5. Scrape or rub away blistered or flaking paint with a wire brush. You can also use a sander or a heat gun to remove tough paint.
6. Use steel wool to remove rust from exposed nail heads. Countersink the nails and fill the hole with filler, then sand smooth and wipe clean before painting.
7. Clean dirt and grime off walls.
8. Clean any moldy areas using the clean-up procedure for molds (see page 24).
9. Choose water-based paint to provide easy clean-up and safer disposal of paint supplies.

How much paint to buy

The quantity of paint you'll need depends on the surface being painted and the type of paint used. A safe estimate is to allow 1 L (about 1 quart) of paint to cover about 8 m^2 (about 86 ft^2) of wall surface for each coat. Several coats may be needed depending on the quality of the paint and the depth of colour being covered.

Apply the paint

Assemble all the tools that you need. Paint only when rain or heavy winds are not expected and when temperatures are predicted to be between 10°C (50°F) and 32°C (90° F). Try to avoid painting in hot, direct sunlight. Always allow the full drying time between coats and remember that temperature and humidity can affect drying times. Follow the instructions on the paint can.

Painting sequence

1. Make sure ladders are secure and cannot slip before using them. Do not place them at too steep an angle. Never step on the top rungs or on an adjacent window ledge. Always try to keep one hand on the ladder. Do not extend your reach.
2. Guard against hazards below. Hang your paint and an empty pail from your ladder so they cannot fall. Use the pail to accommodate rags, brushes and scrapers so you do not have to climb down as often. Keep children and pets away from your work area.
3. Paint the highest parts of your home (for example, soffit boards) first and work your way down. Otherwise, you will get dirt, scrapings and drips in your work.
4. Start painting circular objects like downspouts with diagonal strokes, then finish by painting downward across the strokes. This finish will help prevent the paint from running.
5. Dispose of empty paint containers and dirty rags daily. Put ladders and other equipment away every time you finish work for the day. Seal all partially used paints and store them out of reach of children.

Maintain stucco siding

Skill level rating: 3 - Skilled homeowner

Materials: stucco cement, caulking

Tools: putty knife, small masonry trowel,

Stucco is a strong, durable, economical exterior finish available in a variety of colours. Stucco can be given a new look with paint and holds paint better than a lot of other siding materials because it's more stable at different temperatures.

Maintenance includes gentle cleaning, crack and caulking repair and keeping soil away from the stucco.

1. Hose off stucco gently to remove dirt. Do not use high-pressure washing equipment.
2. Fill and cover hairline cracks with a top application of similarly pigmented cement. Hairline cracks can be caused by natural shrinkage in newly completed stucco. Leave these cracks alone for about two years until shrinkage is finished.
3. Inspect and recaulk where needed around all pipes, intakes and exhaust hoods.
4. Keep soil at least 200 mm (8 in.) below the lower edge of the stucco to prevent water damage to the wall finish, insulation and wall framing.

Maintain masonry

Masonry siding is applied as a veneer and includes brick, stone and their imitations. Masonry should be regularly inspected to spot any problems such as crumbled mortar, cracks or efflorescence. Efflorescence is the formation of white "dust" on masonry. This occurs when salts in the wet mortar dissolve and migrate to the surface. This "dust" is not harmful and will disappear with weathering. Persistence of efflorescence could indicate problems that will need to be repaired including:

- water leakage through mortar
- moisture migrating from inside
- a damaged gutter
- a badly positioned downspout

If you have to clean brick, scrub lightly to avoid damaging the surface or mortar. You can either use a brick cleaning solution found in stores or hire a professional. Commercial brick-cleaners can be caustic (for example, acid), so follow the manufacturer's safety instructions.

Keep soil at least 150 mm (6 in.) below the lower edge of the brick to prevent moisture damage to the wall finish, insulation and wall framing. Other claddings require 200 mm (8 in.) of clearance from the soil.

Skill level rating: 3 - Skilled homeowner

Materials: mortar

Tools: putty knife, small masonry trowel, cold chisel, ladder, smoothing tool, joint tool

Fix masonry cracks

1. Plan the work when freezing temperatures are not expected. Use a cold chisel to chip out any loose mortar from cracks to be filled.
2. Mix a small batch of mortar according to the directions on the package. Use a pointed masonry trowel to apply the mortar to joints.
3. Remove the excess and smooth the mortar using a smoothing tool or tool that will match the type of joint in the surrounding area. A copper tube or wooden dowel produces a smooth, rounded joint.
4. Clean the brick face with a brush and water. Most mortar mixes need to be kept damp for two to three days. Follow the directions on the package.

Maintain flashings and caulking

Cracks around doors and windows can let in dirt, moisture, insects and other pests and add dollars to your heating costs. Repair these cracks and holes as soon as you notice them.

Steel wool is handy for stuffing in larger holes such as service line penetrations. The steel wool serves as a barrier to small rodents and larger insects such as the carpenter ant. Provide an air and moisture seal by inserting low-expansion polyurethane foam into the gap over the steel wool. The foam must be protected from sunlight. Caulk up to a size that is feasible or cover with a flashing material.

Skill level rating: 2 - Handy homeowner

Materials: caulking suited for the job, steel wool, low-expansion polyurethane foam, flashing material

Tools: caulking gun, putty knife, ladder, utility knife, tin snips

1. Check for cracks, gaps and holes that may need caulking.
2. Remove any loose or old material and clean the area.
3. Choose a caulk, foam or flashing that is suitable for the specific job.
4. Apply the steel wool, foam, flashing or caulk with the suitable tools as described above.
5. Quickly smooth the surface of any caulking with a wet finger before the caulking skins over.

ROOFS

The roof covering is essential in protecting your home and its contents from rain, wind, snow and sunlight. Protect your investment through preventive roof maintenance and speedy repairs. Simple maintenance such as trimming tree branches that are close to or touching the roof will prevent damage and discourage insects and moss growth that may affect the durability of your roof. Regular inspections will help you spot any problems such as missing and damaged roofing, popped nails, loose and missing flashing, cracks and gaps in the caulking, holes or rot. Consult CMHC's Homeowner's Inspection Checklist for complete inspection suggestions.

The most common roofing problems are leaks resulting from:

- damaged shingles or flashings
- ice dams
- standing water

Maintenance includes:

- inspecting flashings and caulking
- inspecting and correcting any obstructions of vents in soffits and roof
- inspecting and sealing around roof penetrations such as plumbing vents and chimneys

Prevention tips

- Protect your investment. Conduct regular inspections of the roof (one to two times a year and after a windstorm) to spot trouble early and prevent severe problems. Look for damaged, broken or missing shingles and flashings at the eaves and valleys. Use binoculars to conduct an inspection from the safety of the ground, if possible.
- Regular inspections are also opportunities to note any structural problems such as sags or dips on the roof. Structural problems will need to be assessed by a professional roofing contractor or qualified home inspector.
- When installing Christmas lights, use clips rather than nails to avoid damage to the roof and fascia.

Repair tips

- Purchase roof repair materials at building supply and hardware stores.
- Working on a roof is dangerous. Rent safety equipment from an equipment rental store.
- If in doubt about your repair abilities or if you do not have the proper equipment, call a professional to do the roof work.

Special considerations

Healthy Housing™

- Leaking roofs can result in moisture problems that could encourage premature deterioration, mold growth and possibly cause indoor air quality problems.
- Sealing roof penetrations properly helps to reduce heat loss and energy consumption.

Safety

- Working on a roof is dangerous and requires special equipment and training. Consider hiring a professional roofing contractor to undertake any roofing repairs.

Tasks

Repair leaking roofs

Repair any roof leaks immediately. Most roof leaks begin at the most vulnerable spots—seals along flashing, missing or damaged shingles or tiles, valleys clogged with debris or where standing water at eaves penetrates the sheathing.

When your roof leaks, you must first try to locate the source of the leak. Finding the source is sometimes hard to do because water often travels underneath roofing before it appears as a noticeable leak. If the "leak" shows up as a discolored area on a ceiling or upper part of an outside wall in the spring or during winter thaws, it may not be a roof leak at all. Instead, it may be caused by concealed moisture, which indicates more complex problems in the house.

If you have either extensive leaks or roofing that has deteriorated over several areas, you should hire a reputable roofing contractor. Ask for evaluations and estimates from at least three firms before choosing one for the job.

Skill level rating: 3 - Skilled homeowner

Materials: roof sealant

Tools: caulking gun, putty knife

1. Locate the source of the leak. Look for raised or loose corners on shingles or flashing that may allow water to get under the flashing.
2. Apply roof sealant under the raised edge and press firmly down to seal.

Replace damaged asphalt shingles

Skill level rating: 3 - Skilled homeowner

Materials: replacement roofing or flashing, fibrated roof cement, roofing nails

Tools: flashlight, hammer, pry bar, putty knife, ladder, safety equipment

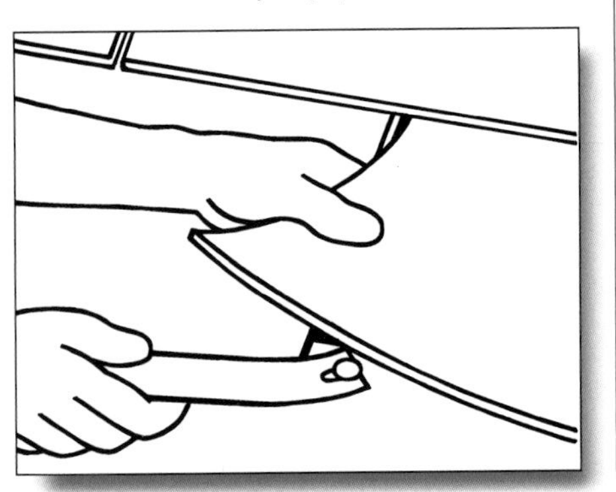

1. This job is best done when the weather is cool. Gently insert a flat pry bar under the upper shingle. Carefully break the seal tab that holds the shingles together. Lift the upper shingle up an inch or two, exposing the nails holding the shingle to be replaced. Remove the damaged shingle from the roof.

2. Insert a replacement shingle.

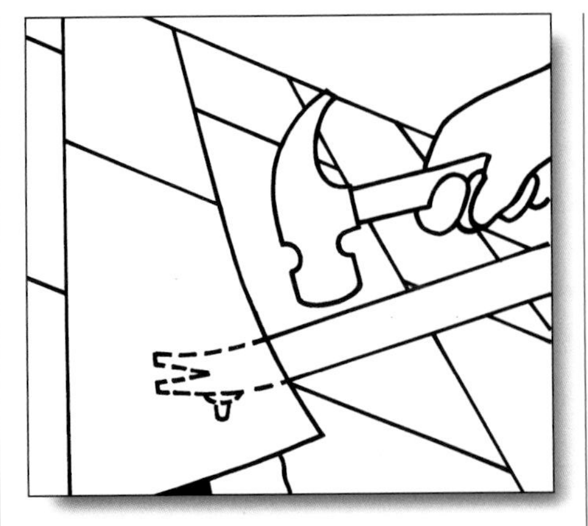

3. If the upper shingle can be bent enough without breaking, nail the new shingle at the top of each tab and at each side. Use galvanized roofing nails. The overlapping shingle should cover the nail heads. If the upper shingle is in danger of breaking, glue the replacement shingle in place using fibrated roof cement.
4. Apply a small amount of roof cement over the nail heads and press the upper shingle firmly in place.

Prevent ice dams

Ice dams form in cold weather when snow on the roof melts (often from lack of sufficient insulation at the outside walls or dormers along with warm air leakage through the ceiling and poor cold air circulation under the roof sheathing). The meltwater then refreezes at the lower edges of the roof. The ice dams back up water, causing it to flow up under the shingles, resulting in leaks in the attic. The solution for ice dams is to improve the insulation and air sealing of the ceiling along with providing an air passage—under the roof sheathing at the eaves so that it stays cold. This job needs to be done by a qualified contractor who will assess your roof and the insulation, airtightness and air circulation conditions in your attic.

Maintain flashings and caulking

Aged or damaged metal flashings can cause roof leaks. Flashings that are worn, rust perforated or damaged will probably need to be replaced. Replacing flashing often involves additional repairs to shingles, replacement of all the shingles and, sometimes, repairs to the roof structure itself. A roofing contractor should do these jobs.

Skill level rating: 3 - Skilled homeowner (repair small holes)
or
Skill level rating: 4 - Qualified tradesperson/contractor (replace flashings)

Materials: roofing cement

Tools: putty knife

1. During the annual roof inspection, note any small holes corrosion or damage to flashings.
2. Fill any small holes with roofing cement.

Maintain vents in soffits and roof

Building codes generally require a 1:300 venting ratio for attics. That is, 10 cm^2 for every 30 m^2 of attic floor area (1 $ft.^2$ for every 300 $ft.^2$). The requirement for low-slope roofs is 1:150. Ideally, that venting would be well distributed, with about half in the soffits and half up higher in the roof in the form of roof spot vents, ridge vents or gable end vents. If soffit vents are to be effective, there needs to be a gap between the underside of the roof deck and the insulation on the ceiling at the edges of the building. That space not only allows the air from the soffits to flow into the attic, it keeps the roof deck cold. When the roof deck is cold, snow melting, ice damming and water leakage are minimized. Commercially available insulation stops can be installed to create that space above the insulation at the eaves.

If you have properly air-sealed the ceiling (see Walls and Ceilings section), you should not need more attic ventilation. Attic ventilation is often overrated. In winter, the cold outside air cannot hold much humidity or carry moisture away from the attic. In summer, attic temperatures are affected more by the sun and shingle colour than by the amount of ventilation. If you wish to improve your attic venting, ensure that it is as well distributed as possible. A combination of soffit vents with either gable end vents, ridge vents or roof spot vents will be sufficient. Turbine vents or other active vent systems won't help and may actually draw more moisture up through a ceiling that is not sealed.

Vents should be screened to keep out birds, animals and insects. Broken or missing vents should be repaired as soon as possible to keep out pigeons and bats. There are serious health issues associated with their droppings including mites that are so small that they can enter easily into the living spaces. For homes that are heavily infested, call a professional to rid the house of these pests and clean up properly.

Note: When repainting soffits, don't paint over the vents and block them.

Bathroom exhaust fan ducts or plumbing stacks may be improperly vented into the attic. Warm, moist air may cause serious moisture, mold or rotting problems. Vent exhaust ducts and plumbing stacks outside (see page 36 in Walls and Ceilings).

Skill level rating: 3 - Skilled homeowner

Materials: pre-made insulation stops or rigid insulation, wood blocks, insect screen, nails

Tools: staple gun, hammer, tape measure, gloves, dust mask, goggles, flashlight, trouble light

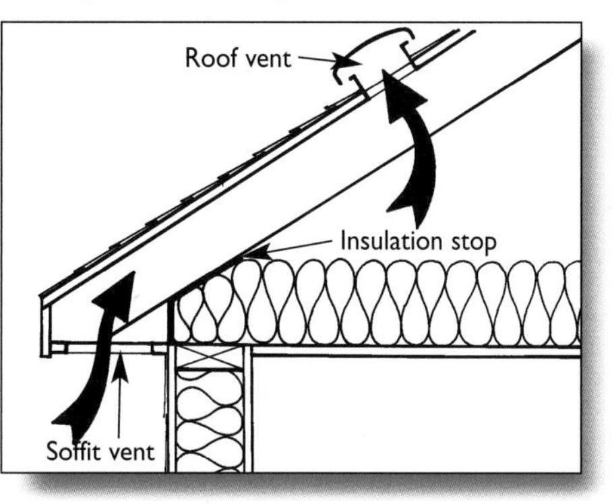

1. Look for soffit venting and other roof venting from the ground. Use binoculars for a better look. Take time to check for loose or missing shingles or any chimney problems.
2. Enter the attic during the day to inspect the venting. Turn off any lights and look for light from the soffit vents. If there isn't light between the insulation and roof deck along the eaves, and you have already confirmed that soffit vents exist, then either the soffit venting must be blocked or the insulation is too tight to the roof deck.
3. Carefully work your way to the eave by stepping only on structural members. Pull back the insulation along the eaves.
4. Install pre-made insulation stops or site-made stops cut from sheets of extruded polystyrene held back from the roof deck by wood blocks nailed to the sides of the rafters. There should be at least a 63 mm (2.5 in.) space between the insulation stop and the roof deck.
5. Refit the insulation so that it covers to the outside edge of the perimeter wall.
6. Ensure that all gable end, ridge vents or roof spot vents are secure and screened.
7. Ensure that all louvred gable end vents are not bent so that they would allow water entry.
8. On the roof, maintain the roofing cement or caulking around roof spot vents to prevent water leaks.

Maintain seals around roof penetrations

Roof penetrations, such as chimneys, skylights, plumbing vents, exhaust fan vents and attic vents, are likely places for leaks to occur. Leaks usually develop when the flashing or caulking around the penetration deteriorates. Sometimes flashings were not installed with adequate overlap or were not correctly positioned. If there are major flashing problems, you may wish to consult with a professional. Otherwise, it may be sufficient to patch existing problems temporarily, maintain the patches regularly and replace the flashing completely when the whole roof is next replaced.

Skill level rating: 3 - Skilled homeowner

Materials: roofing cement (in a can or caulking tube), clear acrylic sealant, foil-backed mastic adhesive tape

Tools: caulking gun, ladder, personal fall protection equipment, putty knife, rag

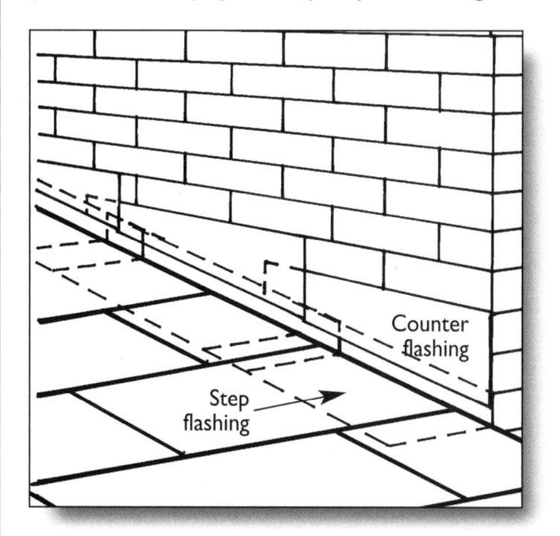

1. Inspect the whole roof, especially any penetrations. Also, check the flashing around dormers or where the roof meets a wall. Each of these penetrations should be flashed. Difficult spots or exposed fasteners should be cemented or caulked.
2. Use roofing cement to dab any exposed fasteners or cement down any shingles as required above flashings. Use roofing cement to

maintain the seal around any vent stacks, as required.

3. Use clear acrylic or siliconized acrylic sealant to caulk the joint where flashings are embedded into brick chimneys (or anywhere else that a clear seal is needed to maintain a good appearance).
4. Temporarily patch any small holes in metal flashings with roofing cement or small pieces of foil-backed mastic adhesive tape.
5. If valley flashings allow water to back up under the shingles, gently pry up shingles beside the flashing and insert a uniform strip of roofing cement. Press the shingles down to form a continuous seal for the length of the valley flashing.

EAVESTROUGHS AND DOWNSPOUTS

Eavestroughs are designed to collect water from your roof. Downspouts drain the water away from your building. Splash blocks are placed at the end of each downspout to help disperse the water onto your lawn. If eavestroughs are not working properly, water spills off the roof and down the walls. This runoff can cause staining, water problems in the basement and peeling paint on the walls. If downspouts are unable to carry away the water, it can cause flooding in the basement and standing water around the foundation and yard.

The most common problems with eavestroughs and downspouts are:

- Built-up debris that prevents proper drainage
- improper adjustment
- minor leaks

Maintenance includes:
- regular removal of debris.

Prevention tips

- Keep eavestroughs clean of debris so they can work well.
- Adjust eavestroughs and downspouts so they drain properly.
- Trim back branches and bushes from the gutter and roof areas

Repair tips

- Use metal screen to prevent leaves and debris from filling and clogging downspouts.

Special considerations

Healthy Housing™

- Eavestroughs and downspouts play an essential role in keeping moisture away from your home. Excessive indoor moisture can lead to mold growth and create an environment for dust mites to grow, potentially causing health problems.
- Attach a rain barrel to the downspouts to collect water for outside watering. This action helps to conserve water and saves on your water bill.

Safety

- Inspecting and maintaining eavestroughs and downspouts involves working on a ladder. Be extremely careful and follow proper safety precautions (see page 13 *Ladders in the Your tool kit* section).
- There are automatic downspout extensions on the market that lower themselves with water flow. These prevent the tripping hazard or breakage from high traffic. Splash blocks still need to be in place.

Tasks

Clean and adjust eavestroughs and downspouts

Skill level rating: 2 - Handy homeowner

Materials: eavestrough straps or spikes depending on your system, screws or nails, silicone sealant, roof cement

Tools: ladder, hose, hammer, screwdriver, plumber's auger

1. Inspect and clean your eavestroughs and downspouts of leaves and other debris every spring and fall. Tighten up any loose joints.
2. Check the outlet where water flows from the eavestroughs into the downspouts. These outlets should have either a leaf guard or a leaf strainer. Clean out each leaf guard or strainer and replace it.
3. Check all eavestrough hangers for tightness. If the hanger is a strap type and is loose, either tighten it with a galvanized screw or replace the nail with a galvanized nail. Replace broken or damaged straps.
4. If the hanger is a sleeve-and-spike type and is loose, nail it again with a galvanized or aluminum spike.
5. Check each downspout for clogs and leaks. If the downspout is clogged, clean it at the eavestrough outlet using a plumber's auger.
6. Check eavestroughs for leaks and proper drainage by pouring water into each eavestrough using a hose or pail. If water does not drain properly, reposition one or more of the hangers furthest away from the outlet until it does. If the hanger is a strap type, you may have to lift the edge of the shingle or other roofing material to expose the strap.
7. Remove the end of the strap from the roof and unscrew or unsnap the attached end of the strap from the eavestrough.
8. Raise the strap to a higher position and secure it again to the roof or fascia board. The new nail or screw hole should be at least 20 mm (3/4 in.) away from the old one.
9. Raise the eavestrough into position and fasten the remaining end of the strap to the gutter.
10. If the hanger is a sleeve-and-spike type, free the eavestrough by pulling the spike out (use a block to support the nail puller) or by cutting the spike with a hacksaw blade.
11. Place another sleeve in another location close by, but at least 20 mm (3/4 in.) from the old location. Raise the eavestrough and refasten it to the fascia board by nailing a new galvanized spike through the sleeve into the board. Cover the nail heads with a dab of roof sealant.

Skill level rating: 2 - Handy homeowner

Materials: silicone sealant, glass fibre patch, roof cement

Tools: ladder, hammer, screwdriver, wire brush, putty knife, caulking gun, cleaning rag

Fix small leaks in eavestroughs and downspouts

1. Locate the leak and remove all debris from the area.
2. Vinyl eavestroughs or downspouts—leaks are usually caused by loose joints or a damaged seal within the joint. Try to tighten the joint. If this doesn't work, replace the part or caulk the leak with silicone sealant.
3. Metal eavestroughs or downspouts—leaks are caused by cracks or holes. Use a wire brush to clean off loose metal and rust.
4. For small leaks, wipe the area clean and spread roof cement over it with a putty knife.
5. Patch leaks larger than 5 mm (3/16 in.) with a small piece of glass fibre patch, 10 mm to 20 mm larger (3/8 in. to 3/4 in.) than the hole. Apply a thin layer of roofing cement over the leak area, place the patch over the cement and press it down firmly. Cover the patch with a heavy coat of cement.

Skill level rating: 1 - Simple maintenance

Materials: splash blocks

Tools: none

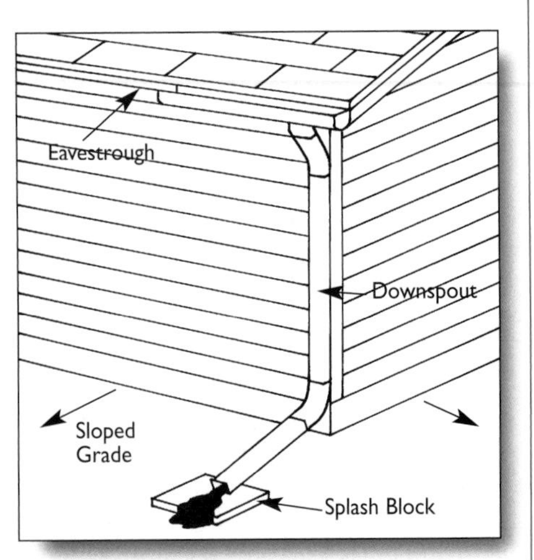

Install splash blocks

1. Place a splash block on the ground under each downspout. The block will direct water away from your home instead of letting it splash on the ground. For the best results, use long splash blocks that will divert the water about 1 m (3 ft.) away, and grade the slope next to your home so it drops about 300 mm (1 ft.) in every 3 m (about 10 ft.).
2. Check the position of the splash blocks each time you inspect and clean out your eavestroughs and downspouts. Splash blocks tend to settle into the ground over time. If you find that they have settled, rebuild the ground underneath them up to its original slope.

STEPS, RAMPS, DECKS AND PORCHES

Steps, ramps, decks and porches require regular maintenance to keep them in good shape and safe.

The most common step, ramp, deck and porch problems are:

- deterioration of wood
- damaged boards
- minor cracks or holes in concrete
- broken concrete

Maintenance includes:

- painting and finishing wood
- inspecting regularly and repairing concrete
- inspecting and tightening railings

Prevention tips

- Broken, chipped, loose or sagging steps are dangerous and unsightly Repair or replace them immediately

Repair tips

- Plan to do your concrete repair when the weather is warm so the concrete will cure properly.
- Use galvanized nails to eliminate rust staining on wood decks.

Special considerations

Healthy Housing™

- Use cedar or redwood as the replacement wood. It has natural preservatives that help it last longer and won't need added chemical preservatives.
- Use a non-toxic ice melter instead of salt to melt ice on concrete walks and steps. These products are easier on the environment and will not cause the concrete to deteriorate.

Safety

- Prevent situations that could cause trips and falls. Keep steps, ramps, decks and porches uncluttered and install railings.
- Repair unevenness or replace any broken concrete sidewalk blocks by lifting them off and levelling and recompacting the base fill.
- Provide good lighting in areas where there are steps. Lights help people to see better and negotiate steps safely.
- Choose nonskid finishes to avoid slips and falls.
- Wheelchair ramps tend to get slippery in winter or wet weather. Install slats at joist locations perpendicular to the ramp combined with a nonslip surface paint and embedded silica sand. Although it is a little harder to keep clean it is safer for people using wheelchairs.

Tasks

Repair or replace wood steps

Damaged or broken wood steps need to be repaired right away.

Skill level rating: 3 - Skilled homeowner

Materials: replacement wood, galvanized nails, finishing nails, paint or stain to match the existing steps

Tools: hammer, framing square, pry bar, paintbrush, portable power circular or crosscut hand saw

Open riser or plank steps

1. Strike the underside of the damaged tread with the hammer until it comes loose. When the nails pop up, pry the tread loose and pull out the nails with the claw of the hammer. . Remove the tread and pull any nails that remain sticking out of the stringer.
2. If required, square one end of the replacement board by aligning the long leg of the square with one of the long edges of the board and marking the board along the short leg of the square. Cut along this line.
3. Measure the length of the tread to be replaced and mark this distance on the replacement board from the squared end inward. Using the square, mark a line at this point and cut the new tread.
4. Put the new tread in place and nail it down. Use at least three nails at each end of the tread, one near each edge of the tread and one near the centre. Use nails that are at least 38 mm (1 1/2 in.) longer than the thickness of the tread.
5. Finish the new tread to match the existing steps.

Closed risers

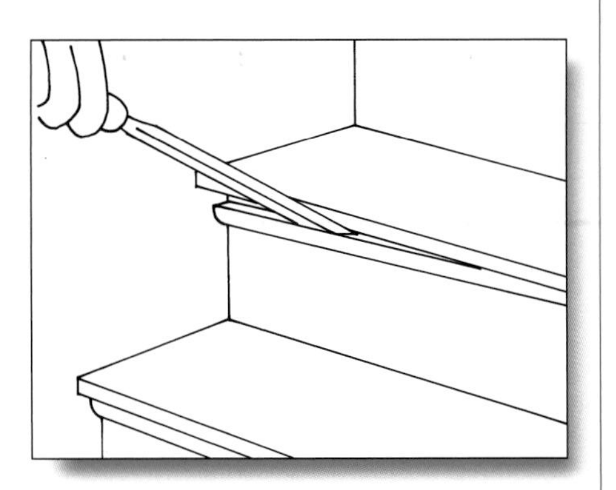

1. The steps may have a piece of trim running along the upper portion of the riser and under the nosing (the part of the tread, if any, that extends beyond the riser). If so, remove it with a pry bar.

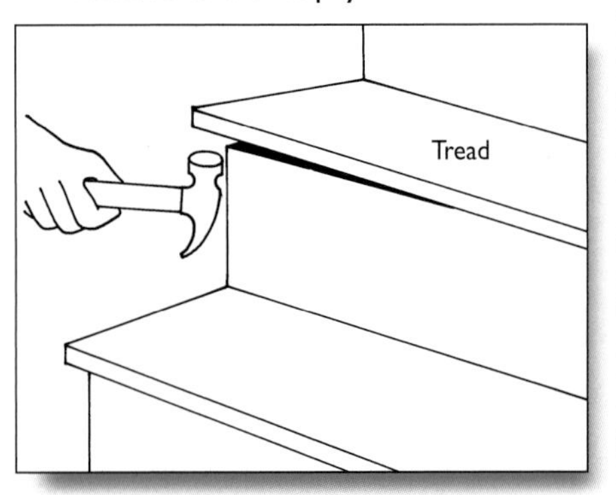

2. If the tread is fastened to the riser, hit the underside of the tread under the nosing to free the tread from the riser below. Use the pry bar to pry up the tread. The tread will also be secured at its sides to the stringers.

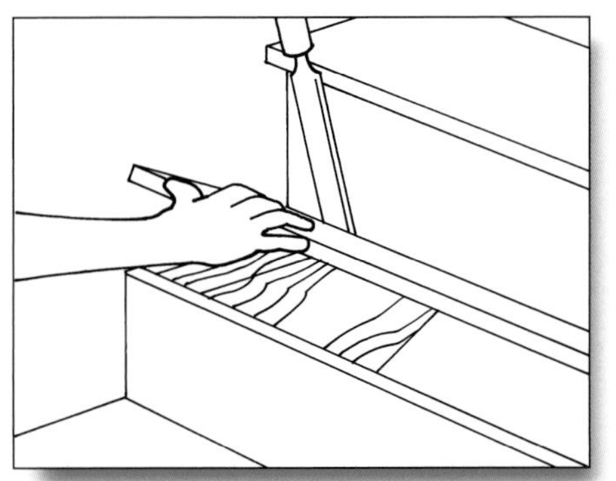

3. After the front and sides of the tread are free, pull the tread forward carefully and evenly to free the back of the tread from the riser above.

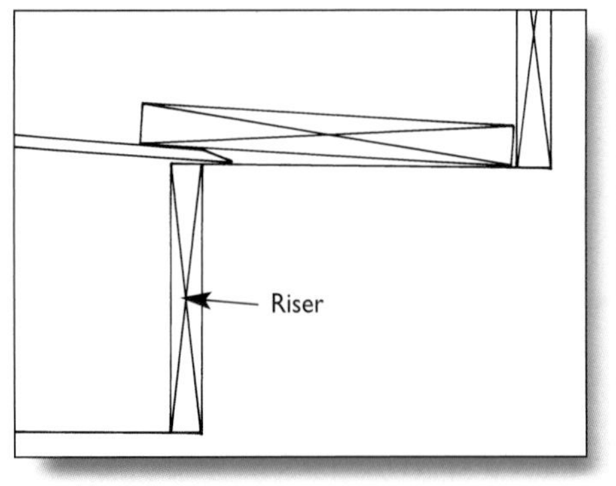

4. Remove the riser if it also has to be replaced. Begin by partially prying the riser loose from the upper tread.

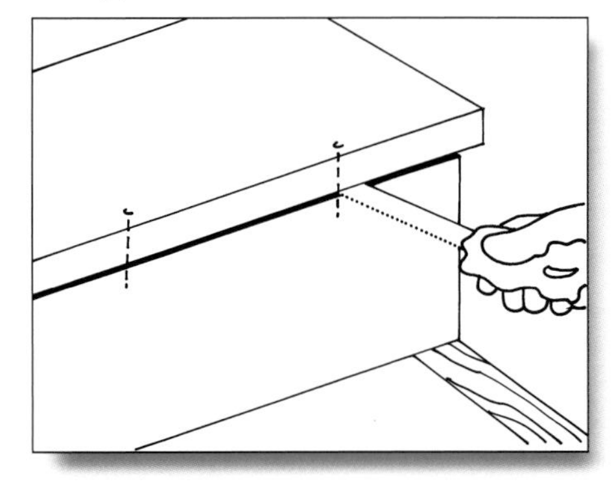

5. Once the nails are exposed, pull them out or saw them off flush with a hacksaw blade.

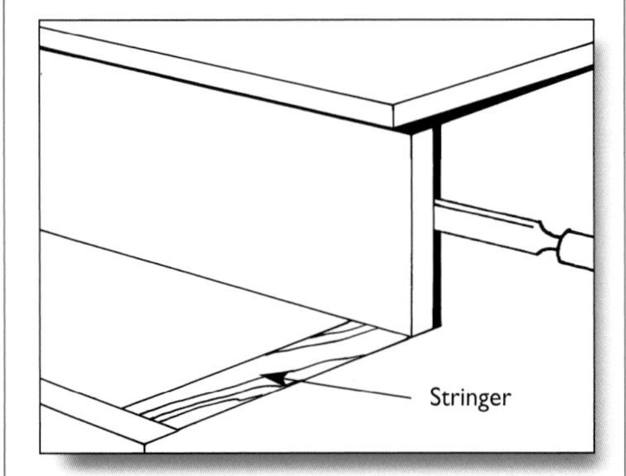

6. Pry the riser away from the stringers. Pull out any nails that are sticking out.
7. Measure and cut the replacement tread and riser.

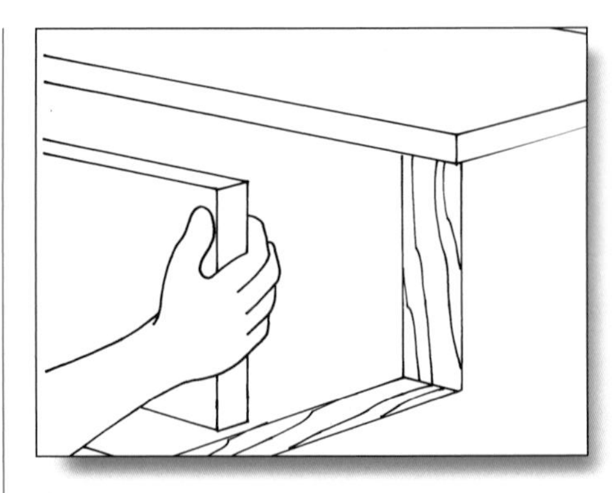

8. Position the new riser under the nosing of the upper tread and up against the cutout of the stringer.

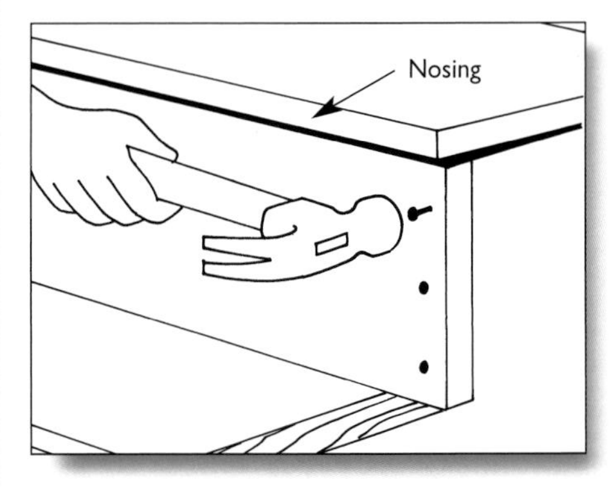

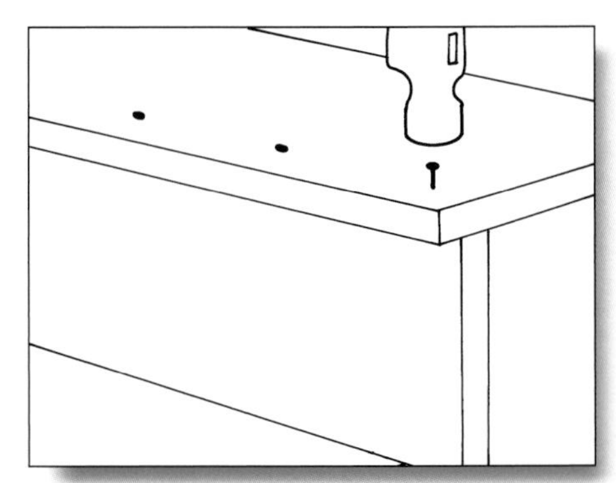

9. Face-nail the riser in place against the stringer. Nail the upper tread down onto the riser.

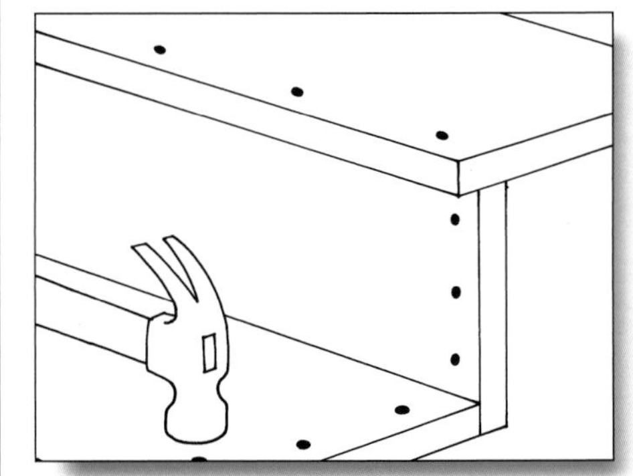

10. Shift the new tread into its final position and face-nail it down onto the stringer and riser below. Space the nails the same as the old nail

spacing or about 150 mm (6 in.) apart. Use at least three nails at each end of the tread, one about 50 mm (2 in.) from each edge and one near the centre.

11. If trim was removed, replace it with small galvanized nails. Finish the new tread and riser to match the existing steps.

Repair minor cracks or holes in concrete

Repairing minor cracks or holes in concrete is easy to do and will help prevent bigger problems. Problems can include surface dusting or scaling, cracks or broken pieces. Flaws in concrete often result from poor original mixing, placing or curing techniques.

Skill level rating: 2 - Handy homeowner

Materials: latex bonding liquid, powdered concrete patcher

Tools: wire brush, chisel, paintbrush, masonry trowel, safety glasses or goggles, breathing protection

1. If the problem is that the concrete surface is dusting or scaling, sweep the surface. Follow with further cleaning and apply the concrete sealer according to instructions given by the sealer manufacturer. CAUTION: sealers commonly contain known carcinogens, so you should use them only outside the house in well-ventilated conditions.
2. To repair any cracks or damaged patches, ensure that all loose concrete is chiselled or scraped out. When chiselling, try to make cracks wider down below than at the surface so patches will lock into place. Clean the area thoroughly.
3. Seal very narrow cracks with sealant or crack filler applied with a caulking gun. For wider cracks, use a concrete patching compound.
4. Specialized patching compounds are also useful for areas where the concrete surface has scaled off or smaller chunks have broken out. Follow the manufacturer's instructions.

Maintain wood with stains and preservatives

Exterior wood steps, ramps, decks and porches need ongoing protection from the sun and weather to maintain durability and an attractive appearance.

Skill level rating: 2 - Handy homeowner

Materials: stain or water-repellent preservative

Tools: brush, rags, drop cloths to protect plants, ladder, protective clothing and eyewear

1. Prepare the surface according to the manufacturer's instructions for the product you are applying. Usually, you will need to clean and lightly sand the surface before applying the product.
2. Use drop cloths to protect surrounding foliage.
3. Apply the stain or preservative following the manufacturer's instructions. Most surfaces require two coats and have to be redone every two to five years. If you can reach it, coat the underside too.

Maintain railings

Although they are outside and often added to a house sometime after the original construction, decks, stairs, handrails and guards must still comply with applicable building code requirements in force at the time of construction. In addition, deck guards and stair railings are exposed to the elements and are subject to weather damage.

Generally, every exterior porch, deck or balcony should be protected by guards on all open sides where the difference in elevation between adjacent levels exceeds 600 mm (24 in.). Guards should be at least 900 mm (35 1/2 in.) high if the deck height is less than 1,800 mm (6 ft.) above the adjacent level. If the deck height is more than 1,800 mm (6 ft.), the guard height increases to a minimum height of 1,070 mm (42 in.). There should be no climbable pieces of the guard between 100 mm (4 in.) and 900 mm (35 1/2 in.) above the deck or balcony floor. Openings through a guard should prevent the passage of a 100 mm (4 in.) sphere. Handrails are required on stairs with more than three risers. Consult the applicable building code for actual requirements.

Skill level rating: 2 - Handy homeowner

Materials: decay-resistant lumber (such as cedar) or compatible deck materials, rust-resistant screws, paint or stain

Tools: measuring tape, saw, drill, screwdriver, sandpaper, sanding block or sander

1. Inspect railings and guards to ensure that they meet the guidelines outlined above and that components and fasteners are secure and in good condition.
2. Upgrade any railings that do not meet the guidelines. This work may require professional assistance.
3. Replace any decayed components.
4. Tighten or replace any loose fasteners.
5. Paint or stain wood regularly to minimize its deterioration due to weather.

GRADING AND DRAINAGE

The lots of most homes are only roughly graded when built. The finished grading and landscaping is usually the responsibility of the owner. Many homes develop flooding and moisture problems because the ground settles, leading to improper grading and poor drainage. The ground surrounding your home should slope away from the house so that surface water is carried away from the foundation.

The most common grading and drainage problems are:

- settlement or inadequate amounts of fill
- poor drainage material

Maintenance includes:

- ensuring proper grade adjacent to the foundation
- preventing erosion
- cleaning and adjusting window wells
- sealing between sidewalks, driveways and walls

Prevention tips

- Make regular inspections to help spot minor problems before they develop into bigger problems.
- Add eavestroughs, if not present on all eaves, along with downspouts, downspout extensions and splash blocks to direct water away from the building.
- Ensure that the grade adjacent to the foundation does not "hold" water as in a flower bed or garden, but directs the surface water away.

Repair tips

- If the grade cannot be built up around the foundation, consider covering the ground with a clay layer. Another option is to bury polyethylene sheeting 150 mm (6 in.), sloped away from the foundation. Either of these methods will help to move water away from the foundation.

Special considerations

Healthy Housing™

- Proper grading and drainage is essential in preventing the entry of moisture into your home. Moisture can lead to premature deterioration of building materials and mold growth, which can have serious health effects.

Safety

- Undertaking major grading and drainage work often requires heavy equipment. Operating this type of equipment requires training and involves safety practices to prevent accidents.

Tasks

Maintain proper grade adjacent to foundation

Extensive grade and drainage work usually requires heavy equipment for digging, spreading, smoothing and hauling earth. It is best to hire a professional contractor to do this work. However, many grading problems are small depressions against the foundation walls where downspouts may direct water. These can often be fixed with the addition of suitable material using a shovel, rake and wheelbarrow.

Skill level rating: 2 - Handy homeowner for minor problems
or
Skill level rating: 4 - Qualified tradesperson/contractor for major problems

Materials: fill, drainage tile, wheelbarrow, shovel, rake

Tools: heavy equipment for hauling and spreading fill

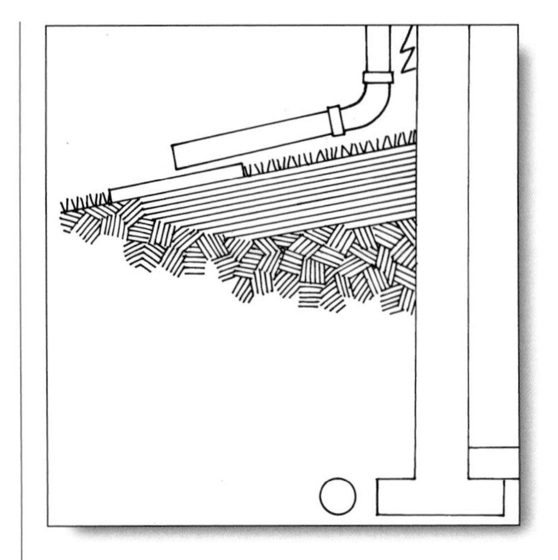

1. The slope should drop away 300 mm (1 ft.) in every 3 m (about 10 ft.). A steeper grade is even better.
2. If your house is on a lot that is level, you should backfill against your foundation as high as possible to about 1.5 m (about 5 ft.) away from the house. Grade down gradually to a slight depression that rises again at your property line. This slope will shed surface water away from the foundation.
3. If your house is on a lot that is lower than the surrounding grade, you may have to use extra fill to allow surface drainage to be directed to street level or to the back of your lot.
4. If the slope around the house is inadequate to provide drainage, you can install in-ground tile drainage systems. You will probably need to hire a contractor to do this job. Another option is to place polyethylene sheeting 150 mm (6 in.) beneath the sod, sloped away from the foundation.

Prevent erosion

Grading and drainage is an essential element in controlling moisture around the foundation. Usually, the backfill placed in the excavation around the foundation settles after the first winter, creating a low spot next to the foundation wall. This low spot allows surface water and roof water to drain down beside the wall where it can possibly overload the perimeter footing drains and leak into the foundation.

The grade around the house should be no higher than 200 mm (8 in.) below the siding. The grade should slope away from the foundation for a distance of about 3 m (10 ft.) at a slope of 1:10 (300 mm drop in 3 m).

Skill level rating: 2 - Handy homeowner for minor problems
or
Skill level rating: 4 - Qualified tradesperson/contractor for major problems

Materials: topsoil, grass seed, sod, downspout extensions, splash blocks

Tools: shovel, garden rake, lawn rake, wheelbarrow

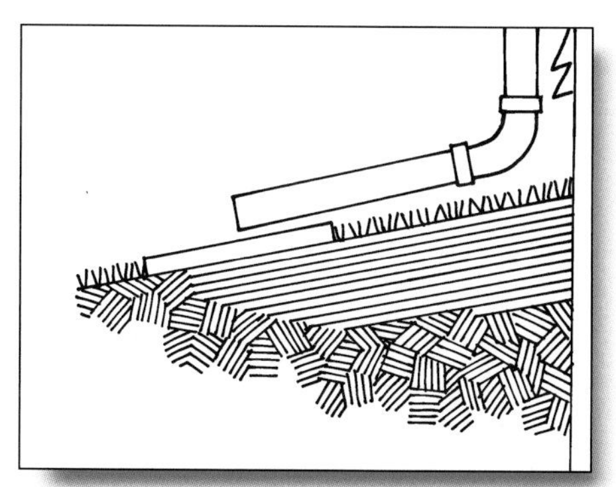

1. Lower any grade adjacent to the foundation that is less than 200 mm (8 in.) below the siding.
2. Slope the grade away from the house with a 300 mm (1 ft.) vertical drop in the 3 m (10 ft.) horizontal distance. If the surrounding grade does not permit this, slope the surface away from the house as much as possible. If necessary, create a shallow swale at least 3 m (10 ft.) from the house so that at least the grade adjacent to the house does not slope toward the foundation. Ensure that surface run-off from nearby areas drains away from the house along the swales.
3. Finish the grading with a layer of top-soil or clay that will provide a cap to reduce surface water penetration next to the foundation.
4. Plant grass seed on the newly graded slope. Using sod provides more immediate erosion control at a higher price.
5. If there aren't eavestroughs along all eaves, install eavestroughs along with downspouts, downspout extensions and splash blocks to divert roof water away from the foundation.

Clean and adjust window wells

Windows or parts of windows that are below grade should be protected by window wells. The bottom of the well should be 200 mm (8 in.) minimum below the bottom edge of the window. The bottom should consist of free-draining gravel to permit good drainage toward the footing drain and to prevent water pooling in the window well. If the backfill doesn't drain well, a column of gravel can be used to connect the bottom of the window well with the perimeter drain tile if one exists.

The slope around a window well should direct surface water away. Ensure that all water, especially the spring melt water from the roof, flows away from the building and not into the window well. If the water freezes in the window well during the night, the next day's melt may flow over the ice, through the window and into the basement.

Walls of window wells are usually made of corrugated metal or plastic. Although they may be fastened to the foundation wall, in some soils frost action pushing up the window well may damage the foundation wall. It is often better to place the window well in position and adjust it every few years to correct displacement due to frost action.

Skill level rating: 2 - Handy homeowner

Materials: crushed stone or gravel

Tools: shovel, level, tape measure, garden rake, lawn rake

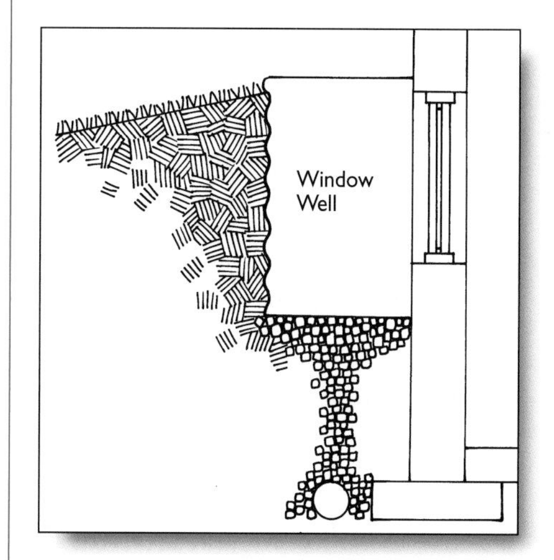

1. Periodically remove any leaves or debris from window wells.
2. Ensure that the bottom of the window well is low enough below the window and is free draining.
3. If the window well has been displaced by frost action or settling, remove it. Use a layer of gravel or crushed stone to level the bottom. Set the window well back in place so that the top is level and at the correct height. Replace the backfill. Slope away from the house and window well. Provide a clay cap on the surrounding grade as described above.

Seal cracks between sidewalks, driveways and walls

Sidewalks and driveways that are next to a foundation wall can provide a good cap to direct surface water away from the foundation. However, some sidewalks and driveways may be either flat or may actually slope toward the wall due to settling of backfill. Consult a contractor about correcting a severe sloping problem.

If the sidewalk or driveway is level or sloping away from the foundation, it is still necessary to maintain a seal between those surfaces and the foundation wall to prevent water from leaking into that crack.

Skill level rating: 2 - Handy homeowner

Materials: polyurethane sealant (tubes), driveway crack filler (tubes)

Tools: wire brush, caulking gun

1. Clean any loose debris or old caulking out of the crack with a wire brush.
2. Fill deep cracks to about 13 mm (1/2 in.) below the surface with sand.
3. Seal the crack with polyurethane caulking for either sidewalks or pavement, or with driveway crack filler for paved surfaces.

Getting More Help

Getting More Help

Maintenance and repair jobs require a variety of skills and training. Not every homeowner has the time or skills to do every job around the house. Luckily, there are qualified professionals who can give you the help that you need.

Tips for Hiring a Professional

- Decide what type of professional you need to hire. If you need someone to inspect your home to find the source of problems, you can call a professional home inspector. If you have a specific problem with a part of your house, you'd need to call a specialist in that field. For example, a problem with your furnace would require a qualified furnace technician. If the problem is related to poor indoor air quality, you can call an indoor air quality professional who has completed the CMHC Residential IAQ Investigator Program. (Your local CMHC office keeps a listing of people in your area who've completed the Residential IAQ Investigator Program).
- Get referrals from family, friends and neighbours.
- Choose someone with experience and who is a member of a professional association for their industry. Depending upon the industry and the province, the professional may need to be licensed or registered to legally carry out their work.
- Get estimates from at least three contractors.
- Get the estimate in writing. For any work, getting a written agreement that explains the details about the job and everyone's responsibilities is essential. In many cases, such as home inspections or indoor air quality inspections, you should expect a final written report. You'll need this important documentation for reference as you proceed with further repairs.

For a full checklist of all the steps included in hiring a contractor or a home inspector, refer to CMHC's *About Your House* factsheets, *Hiring a Contractor*, 62277 and *Hiring a Home Inspector,* 62839.

Learn More

CMHC has many publications that can help you learn more about caring for your home.

To order these publications and to find out about other CMHC products, contact:

Your local CMHC office
or
Canada Mortgage and Housing Corporation
700 Montréal Road
Ottawa ON K1A 0P7
Phone: 1-800-668-2642
Fax: 1-800-245-9274

Visit our home page at
www.cmhc.gc.ca

Suggested Publications

Title	CMHC publication number
Before You Renovate Renovation Guide and Catalogue	61001
Building Materials for the Environmentally Hypersensitive	61089
Canadian Wood-Frame House Construction	61010
Clean up Procedures for Mold	61091
Glossary of Housing Terms	60939
Healthy Housing Renovation Planner	60957
Homeowner's Inspection Checklist	62114
Homeowner's Manual	61841
Household Guide to Water Efficiency	61924
Renovator's Technical Guide	61946
The Clean Air Guide: How to Identify and Correct Indoor Air Problems in Your Home	61082
About Your House fact sheets—free	
After the Flood	60515
Attic Venting, Attic Moisture, and Ice Dams	62034
Carbon Monoxide	62046
Choosing a Dehumidifier	62045
Combustion Gases in Your Home	62028
Fighting Mold	60516
Hiring a Contractor	62277
Hiring a Home Inspector	62839
Maintaining Your HRV	62043
Measuring Humidity in Your Home	62027
Removing Ice on Roofs	62036
Sample Renovation Contract	62351
Should You Get Your Heating Ducts Cleaned?	62044
Testing Airflow	62288
The Importance of Bathroom and Kitchen Fans	62037
Your Furnace Filter	62041
Your Septic System	62795
Before You Start:	
A New Addition	62268
An Energy Efficient Retrofit—The Building Envelope	62264
An Energy Efficient Retrofit—Mechanical Systems	62262
Assessing the Comfort and Safety of Your Home's Mechanical Systems	62266
Assessing the Renovation Project	62246
Renovating Your Bathroom	62254
Renovating Your Basement—Moisture Problems	62250
Renovating Your Basement—Structural Issues and Soil Conditions	62248
Renovating Your Kitchen	62252
Repairing and Replacing Materials—Exterior Walls	62260
Window and Door Renovations	62256
Repairing or Replacing Roof Finishes	62258

Glossary

Glossary

ABS–***(ABS)*** Acronym for acrylonitrile-butadiene-styrene. A type of rigid plastic used in plumbing pipes for drain, waste and vent systems. Can also be used for potable water.

Air barrier–Material incorporated into the house envelope to retard the movement of air. Called air-vapour barrier when it retards air and moisture.

Air sealing–The application of weather stripping such as caulking and expanding foam to close off small cracks and spaces at windows and doors and on walls and ceilings to reduce air leakage and consequent heat loss.

Ampere–The unit of electrical current equivalent to the steady current produced by one volt applied across a resistance of one ohm.

Backdrafting (flow reversal)–The reverse flow of outdoor air into a building through the barometric damper, draft hood or burner unit as a result of chimney blockage or a pressure differential greater than can be drawn by the chimney. Backdrafting causes smell, smoke or toxic gases to escape into the interior of a building. "Cold" backdrafting occurs when the chimney is acting as an air inlet but there is no burner operation or just a smouldering fire in a fireplace. "Hot" backdrafting occurs when the hot flue gas products are prevented from exhausting by flue reversal.

Barometric damper or barometric draft regulator–A device which functions to maintain a desired draft in the appliance by automatically reducing excess chimney draft to the desired level.

Baseboard heater–A radiator shaped like a baseboard having openings at top and bottom through which air circulates.

Brads–Thin nails with a small head, used for small finish panel-moulding and so on.

Bridging–A method used to resist twisting of joists and for stiffening floor construction by fitting either crossed pieces or solid blocks between the joists.

Cap flashing–Sheet metal or other material used above a window or door to shed water.

Caulk–To make tight with a sealing material.

Caulking–Material with widely different chemical compositions used to make a seam or joint air-tight or watertight.

Ceramic tile–Decorative ceramic tiles of various shapes and size, normally used where durability is important and excessive exposure to moisture could occur.

Combustion appliance–A fuel-burning heating or cooking appliance such as an oil or gas furnace, wood burning stove, oil or gas space heater or a gas range.

Condensation–The transformation of the vapour content of the air into water on cold surfaces.

Convector–A heating device in which the air enters through an opening near the floor, is heated as it passes through the heating element and enters the room through an upper opening.

Cross-bridging–Small wood planks or metal pegs that are inserted diagonally between adjacent floor or roof joists.

Dampproofing–(1) The process of coating the outside of a foundation wall with a special preparation to resist passage of moisture through the wall. (2) Material used to resist the passage of moisture through concrete floor slabs and from masonry to wood.

Downspout–A pipe, which carries water from the eavestrough to the ground or the storm drainage system.

Eave–The part of the roof that projects beyond the face of the wall.

Eavestrough–A trough fixed to an eave to collect and carry away the run-off from the roof. Also called gutter.

Efflorescence–Formation of a white crystalline deposit on the face of masonry walls.

Electrolysis–An electrochemical reaction between two dissimilar metals, such as copper and galvanized steel, causing corrosion of a joint where the two materials are in contact with each other.

End nailing–Fastening two pieces of lumber together by nailing through the surface of one piece and into the end grain of the other. For example, solid floor framing blocking can be end-nailed by nailing through the side of the joists and into the end grain of the block.

Entrained air–Trapped air bubbles.

Fascia–A finish board around the face of eaves and roof projections.

Fixed-pin hinge–A type of hinge in which the hinge pin is not removable.

Flashing–Sheet metal or other material used in roof and wall construction to shed water.

Forced air–Air circulated through ductwork within a house by means of a circulating fan located in the furnace housing.

Fuse–A device capable of automatically opening an electric circuit under predetermined overload or short-circuit conditions by fusing or melting; an overcurrent device.

Gable end vents–A sheet metal or plastic vent in the end of a gable or dormer roof on a house.

Glazing–A generic term for the transparent, or sometimes translucent, material in a window or door. Often, but not always, glass.

Ground fault circuit interrupter–A device designed to interrupt, almost instantaneously, an accidental connection between a live part of an electrical system and ground (a short-circuit or a shock) when the current exceeds a very small predetermined value. This device will react to a dangerous situation before a fuse or circuit breaker, and before a person can be harmed by the shock.

Grout–A thin mixture of cement mortar and additional water.

HRV–Acronym for Heat Recovery Ventilator. A ventilation system that provides fresh outdoor air to the house while extracting heat from the stale outgoing air. HRVs help keep indoor humidity levels under control, improve indoor air quality and may keep heating costs down.

Hydraulic cement patch–A cement material which will harden under water. Quick setting, hydraulic cement patching materials are used to quickly patch small water leaks in concrete structures.

Hygrometer–An instrument designed to measure the relative humidity of the atmosphere.

IAQ–Acronym for Indoor Air Quality. A general term relating to the presence of chemical and biological contaminants in the air within a building, and their potential health effects.

Joist–One of a series of horizontal wood members, usually 50 mm (2 in.) nominal thickness, used for support in floors, ceilings or roofs.

Kerf–A groove or cut.

Load-bearing–Subjected to or designed to carry loads in addition to its own weight (as applying to a building element).

Lockset–A doorknob assembly that includes knobs or lever handles, a latch bolt assembly and trim.

Loose-pin hinge–A type of hinge in which the pin can be pulled from the barrel so that the two hinge leaves separate.

Marrette–A type of twist-on wire connector used to secure and protect the twisted ends of two joined wires.

Mastic–Any of various pasty materials used as a protective coating.

Moisture barrier–Any material which is used to retard the passage or flow of vapour or moisture into construction and prevent condensation.

Mortise–The cut-out in a board or unit to receive a tenon lock, hinge and so on.

Muntin–A horizontal member which divides panes of glass, windows, or doors.

Nosing–The rounded and projecting edge of a stair tread, window, sill and so on.

Oriented strand board, or OSB–Structural wood panel manufactured from wood strands that are oriented in the same direction and bonded together with glue. It is a high strength product made from low-grade (waste) material.

Parging–A coat of plaster or cement mortar applied to masonry or concrete walls.

Passage set–Doorknobs or levers, latch bolt assembly and trim that does not have any locking mechanism.

Pressure tank–A water supply holding tank in which the incoming water pushes a cushion of air to the top of the tank until it reaches a pre-set pressure, causing the pump to shut off. The pressure in the tank allows water to be drawn off until the pressure reaches a lower pre-set level and the pump comes on. The pressure tank allows water to be used without the pump coming on each time.

Radiator–That part of the system, exposed or concealed, from which heat is radiated to a room or other space within the building; heat transferring device.

Ridge vent–A special sheet metal or plastic vent which is installed along the ridge of the roof.

Riser–(1) The vertical board under the tread in stairs. (2) In plumbing, a supply pipe that extends through at least one full storey to convey water.

Roof boot–A pre-formed rubber flashing that fits around a plumbing vent stack to be integrated with the roofing to provide a water-tight seal.

Roof spot vent–A roof vent designed to be installed on the surface of the roof and integrated with the roofing to provide a water tight seal while allowing the passage of air.

Rose–The wide flat part of a doorknob that fits snugly against the door.

Sacrificial anode–A rod, made of magnesium or aluminum, which is wrapped around a steel core wire that is screwed into the top of a hot water heater tank to prevent the tank from rusting.

Sash–A light frame of wood, metal, or plastic either fixed or movable, which holds the glass in a window.

Sealant–A flexible material used on the inside (or outside) of a building to seal gaps in the building envelope to prevent uncontrolled air infiltration and exfiltration.

Self-closure device–A pneumatic or hydraulic door closing device designed to close the door automatically.

Shim–A thin piece of material (sometimes tapered) used to fill in space between objects.

Silicone sealant–A solvent-free silicone compound that is highly durable and excellent for sealing large moving joints. Ventilation is required during application and curing.

Soffits–The underside of elements of a building such as staircases, roof overhangs, beams and so on.

Solder–A metallic compound that becomes liquid when heated and is applied in the liquid state. Used to join copper plumbing pipes and fittings.

Solvent–A substance, usually liquid, having the power of dissolving the base material of a paint.

Spline–A rectangular strip of wood that is substituted for the tongue fitted into the grooves of two adjoining members. In window screens, it is the thin strip that holds the screen into the frame

Stringer–(1) A long, heavy horizontal timber that connects upright posts in a structure and supports a floor. (2) The inclined member that supports the treads and risers of a stair

Stucco–Any cement-like material used as an exterior covering for walls and the like; it is put on wet and dries hard and durable.

Stud finder–An electronic device used to detect changes in wall density or a magnetic device used to locate hidden nails or screws. Either way, structural support present in walls or ceilings can be detected.

Subfloor–Boards or sheet material laid on joists to support the finished floor.

Sump–A watertight tank that receives the discharge of drainage water from a subdrain or a foundation drain and from which the discharge is ejected into drainage piping by pumping.

Sump pump–A pump, usually electrically operated, to remove water that collects in a sump.

Swale–A small landscaping channel that is usually grassed and is wider than deep.

Tread–The horizontal part of a step, as opposed to the vertical riser.

Turbine vent–A roof spot vent with a mushroom shaped top composed of wind vanes that pull attic air up through the roof (and possibly pull warm, moist air up into the attic from the living space).

VOC–Acronym for Volatile Organic Compound. One of a group of organic chemicals that can be a gas or vapour at indoor temperature. They are found in many common products such as oil-based paints and varnishes, caulking, glues, synthetic carpeting and vinyl flooring and so on. They contribute to poor indoor air quality.

Water vapour–Water in a gaseous state and present in the atmosphere in varying amounts.

Waterproofing–Control of liquid water to prevent it from entering a building. Regarding foundations, "dampproofing" is designed to control soil moisture in the form of capillary water or water vapour, whereas "waterproofing" is designed to resist ground water in situations where the water table in the vicinity of the foundation may rise higher than the foundation floor with no guarantee that it will drain away successfully.

Index

– A –

– B –

– C –

– D –

– E –

– F –

– G –